LABORATORY MANUAL

CLARINGTON'S HUMAN ANATOMY & PHYSIOLOGY (I)

[FOR ONLINE STUDENTS]

PART ONE

FIRST EDITION

DR. MORRIS B. CLARINGTON

Published by

Conceptual Integrative Solutions Global, LLC
"Gearing Minds for Tomorrow's World"

NOTE FROM AUTHOR

This laboratory manual has been assembled to provide students with an understanding to Biology from a Human Anatomy and Physiology perspective. Students often fell to see the connection between a strong foundation in the Natural Sciences and their success in prospective Healthcare Occupational courses. My goal was to bridge the gap between the Natural Sciences and Healthcare Occupational courses.

Your journey as a healthcare provider begins with you acquiring the necessary skills and knowledge to be proficient in your specialty. Your dedication, compassion, knowledge, attention to detail, and service will one day provide comfort to patients, and offer your life purpose. It is my belief that there is no greater sense of accomplishment than committing one's self to the service and well-being of others.

© 2017 by Dr. Morris B. Clarington. All rights reserved.

Terms of Use Policy

Labs in this manual were compiled from various sources which include free animated, interactive websites on anatomy and physiology, free websites by educators, schools, governments, & nonprofits, and free information offered by commercial websites. Most of the resources are public domain resources and open educational resources. Public domain is a designation for content that is not protected by any copyright law or other restriction and may be freely copied, shared, altered and republished by anyone. Open educational resources (OER) are freely accessible, openly licensed text, media, and other digital assets that are useful for teaching, learning, and assessing as well as for research purposes.

The open educational resources (OER) website hyperlinks to other internet sites featured in this lab manual are listed for your convenience. These other sites are maintained by third parties over which Conceptual Integrative Solutions Global nor your learning institution (school, college, university, etc.) exercises any control. Once you enter these website hyperlinks, whether through an advertisement, another service or content link or through a new browser window, be aware that Conceptual Integrative Solutions Global nor your learning institution is responsible for the privacy practices of these other sites.

The Open educational website hyperlinks may feature materials, information, products, and services provided by third parties. This inclusion of advertisements on these websites does not imply Conceptual Integrative Solutions Global nor your learning institution's endorsement of the advertised products or services. Conceptual Integrative Solutions Global nor the learning institution shall be responsible for any loss or damage of any kind incurred as a result of the presence of such advertisements on these websites. You shall be solely responsible for any correspondence or transactions you have with any third party advertisers.

Every effort has been made to provide you with longstanding, safe, creditable, and reputable, website hyperlinks. If you find that a particular website hyperlink does not work, please notify your instructor so that they may provide you with an alternative website hyperlink that appropriately reflect the concept being taught. We encourage you to review the privacy statements of each website that you visit. If you decide to access any of the third party sites linked, you do so entirely at your own risk.

Some of the materials in this lab manual are original exercises from Conceptual Integrative Solutions Global, and are subject to copyright laws; however, all labs are common to most general biology courses but have been modified to fit the healthcare theme of this manual. No part of this publication may be reproduced, distributed, or transmitted in any form or by any means, including photocopying, recording, or other electronic or mechanical methods, without the prior written permission of the publisher or author. The exception would be in the case of brief quotations embodied in the critical articles or reviews and pages where permission is specifically granted by the publisher or author.

Although every precaution has been taken to verify the accuracy of the information contained herein, the author and publisher assume no responsibility for any errors or omissions. No liability is assumed for damages that may result from the use of information contained within.

Hyperlink policy

This laboratory manual provides hyperlinks to other locations or websites on the Internet. These hyperlinks lead to websites published or operated by third parties who are not affiliated with or in any way related to **Conceptual Integrative Solutions Global or your learning institution (school, college, university, etc.).** They have been included in our manual to enhance your user experience and are presented for information purposes only. We endeavor to select reputable websites and sources of information for your convenience.

However, by providing hyperlinks to an external website or webpage, **Conceptual Integrative Solutions Global nor your learning institution** shall be deemed to endorse, recommend, approve, guarantee or introduce any third parties or the services/ products they provide on their websites, or to have any form of co-operation with such third parties and websites unless otherwise stated.

We are not in any way responsible for the content of any externally linked website or webpage. You use or follow these hyperlinks at your own risk and **Conceptual Integrative Solutions Global nor is your learning institution** responsible for any damages or losses incurred or suffered by you arising out of or in connection with your use of the link. **Conceptual Integrative Solutions Global nor your learning institution** is not a party to any contractual arrangements entered into between you and the provider of the external website unless otherwise expressly specified or agreed to by **Conceptual Integrative Solutions Global or your learning institution**.

Any hyperlinks to websites that contain downloadable software are provided for your convenience only. Again we are not responsible for any difficulties you may encounter in downloading the software or for any consequences from your doing so. Please remember that the use of any software downloaded from the Internet may be governed by a license agreement and your failure to observe the terms of such license agreement may result in an infringement of intellectual property rights of the relevant software provider which we are not in any way responsible.

Please be mindful that when you click on a link and leave our website you will be subject to the terms of use and privacy policies of the other website that you are going to visit.

No Warranties

While every care has been taken in preparing the information and materials contained in this site, such information and materials are provided to you "as is" without warranty of any kind either express or implied. In particular, no warranty regarding non-infringement, security, accuracy, fitness for a particular purpose or freedom from computer virus is given in conjunction with such information and materials.

IMPORTANT: By using this laboratory manual and any of its pages you are agreeing to the terms set out above.

Printed in the United States of America
First Printing, 2017
ISBN-13: 978-1978372672
ISBN-10: 1978372671
For Questions Contact:
Conceptual Integrative Solutions Global, LLC
1-912-495-8745

www.weseesuccess.com

PREFACE

This lab manual was also created to be an affordable option to colleges and/or universities that desire to provide an experimental foundation for the theoretical concepts introduced in the lectures. Labs are intended to reinforce those concepts discussed in lecture.

Many of the labs used are virtual labs. Virtual Labs are web sites or computer software for interactive learning based on simulation of real phenomena. It allows students to explore topics by comparing and contrasting different scenarios, to pause and restart application for reflection and note taking, to get practical experimentation experience over the Internet. When compared to traditional laboratories, virtual laboratories are useful when some experiments involve expensive equipment or when chemicals may cause harm to human health.

For those exercises that are not in virtual space, and require you to purchase inexpensive items, most of the materials used in the experimentation process for this manual can be purchased from most local variety stores.

Labs in this manual are divided into several exercises. This allows an instructor to select those exercises that will best meet their needs. The laboratory manual is designed for students with minimal backgrounds in the physical and biological sciences who are pursuing careers in the healthcare profession. Labs will encompass: Laboratory Safety, Body Organization, Chemical Basis of Life, Cell Structure and Functions, Tissue Classifications, Integumentary System, Skeletal System, Muscular System, and Nervous Sensory Systems.

Each lab is arranged in three sections: **CRASHCOURSE VIDEO(S)**, **DEFINING KEY TERMS**, and **LAB EXERCISES**.

- **CRASHCOURSE VIDEO(S)** is an educational YouTube channel. The host provides a fast, extensive overview of important anatomical and physiological concepts in a brief 10 to 20 minute video. (Please watch the video link.) **https://youtu.be/hIowxzmCDpw**

- **DEFINING KEY TERMS** refer to important words or expressions that are essential to the student's understanding of the course material or study concept. A definition is a statement expressing the essential nature of something. Definitions enable us to have a common understanding of a word or topic; they allow us to all be on the same page when discussing or reading about a concept. Each word or expression and its definition should be reviewed repeatedly to reinforce rote learning (a memorization technique based on repetition).

- **LAB EXERCISES** are designed to provide students with an experimental foundation for the theoretical concepts introduced in the lectures.

The Open educational resources (OER) used for this book can be collaborated with any of the major online/ distance learning platforms such as: Blackboard, Moodle, or Desire2Learn (D2L).
Please make sure that your computer software is up-to-date.

Computer Requirements:
- **Laptop computer with a minimum of a 2GHz processor and 2GB of RAM**
- **Desktop computer with a minimum of a 2GHz processor and 4GB of RAM**
- **A DSL, cable connection, or greater. (Dial-up is not sufficient)**

- Windows 7 or newer system operating software for PC computers
- Disable pop-up blockers
- Antivirus
- Disable all third-party toolbars
- Microsoft Office Suite (Word, Excel, PowerPoint)
- Update your web browser. (Try using Mozilla Firefox, Google Chrome, or Safari)
- Java (Latest Java version)
- Adobe Acrobat Reader
- Adobe Flash Player
- Adobe Shockwave Player
- RealPlayer
- Windows: Windows Media Player or Mac: QuickTime
- Audio

Please Note: Your instructor may have the web addresses added as embed codes or hyperlinks that have been integrated into your college's preferred online platform. If this is the case, simply click on the icon or web link to begin the exercises. ONCE YOU ARRIVE TO THE DESIRED WEBSITE ADDRESS, PLEASE DO NOT CLICK ON ANY OF THE THIRD PARTY ADVERTISEMENTS. IF YOU DO SO, YOU MAYBE ROUTED AWAY FROM THE DESIRED WEBSITE.

You will be required to purchase the following items to complete the various exercises in this laboratory manual. Most of the items can be purchased from your local dollar store, if you do not have them in your home already. Items such as Reflex Hammer, C512 Tuning Fork, and Penlight may have to be purchased online. Please do not purchase the expensive or name-brand versions of any of these items. This will cut down on your material cost.

- 1 pack of Cotton Swabs
- 1 pack of Pencils
- 1 roll of Adhesive Tape
- 1 roll of Clear Tape
- 1 roll of Masking Tape
- 1 roll of Paper Towels
- 1 standard-sized 18" x 26" x 1" Aluminum Bun Pan/Sheet Pan
- 10 count pack of Washable Markers
- 12 count pack of Color Pencils
- 2 Coins (*Pennies*)
- 2 quarts of White Distilled Vinegar
- 2 tablespoons of Corn Syrup
- 5 bendable drinking straws
- 6 chicken bones (*can be from cooked specimen*)
- 6 mason jars
- 8 to 10 count pack of Disposable Gloves
- Apron
- C512 Tuning Fork
- Chair
- Clear drinking glass (*with a clear bottom*)
- Forceps or (Tweeters)
- Goggles/Safety Glasses
- Iodine

- Millimeter ruler
- One 16 oz. box of Cornstarch
- One 20oz - 32oz bottle of liquid dish soap
- One 2-Liter Coke
- One 44 ounce cup
- One gallon of Liquid Bleach
- Pack of White Copy Paper
- Pen Light
- Reflex Hammer
- Spoon
- Tap Water
- Three 16oz bottles of 70% Isopropyl Alcohol
- Timer or Stopwatch (*you can use the timer on your smartphone*)

TABLE OF CONTENTS

LABORATORY SAFETY RULES & PROCEDURES … p. 10

LAB 1 -- LABORATORY SAFETY … p. 14

LAB 2 -- BODY ORGANIZATION … p. 32

LAB 3 -- CHEMICAL BASIS OF LIFE … p. 42

LAB 4 -- CELL STRUCTURE AND FUNCTIONS … p. 52

LAB 5 -- TISSUE CLASSIFICATIONS … p. 79

LAB 6 -- THE INTEGUMENTARY SYSTEM … p. 89

LAB 7 -- THE SKELETAL SYSTEM … p. 103

LAB 8 -- THE MUSCULAR SYSTEM … p. 146

LAB 9 -- THE NERVOUS AND SENSORY SYSTEMS … p. 162

LABORATORY SAFETY RULES & PROCEDURES:

<u>All students must read and understand the information in this section with regard to laboratory safety and emergency procedures prior to the first laboratory session.</u> Your personal laboratory safety depends mostly on you. An effort has been made to address situations that may pose a hazard in the laboratory setting, but the information and instructions provided cannot be considered all-inclusive.

Good common sense is needed for safety in a laboratory. With good judgment, the chance of an accident is very small. Nevertheless, healthcare facilities are full of potential hazards that can cause serious injury and or damage to the equipment.

It is expected that each student will work in a responsible manner and exercise common sense and good judgement. If at any time you are not sure how to handle a particular situation, ask your Instructor for advice. Do not touch anything with which you are not completely familiar. It is always better to ask questions than to risk harm to yourself or damage to the equipment.

<u>ALTHOUGH YOUR LABORATORY EXPERIENCE WILL BE CONFINE TO THE ONLINE LEARNING PLATFORM AND YOUR HOME, IT'S STILL IMPORTANT TO KNOW STANDARD LABORATORY SAFETY GUIDELINES AND PROCEDURES.</u>

General Laboratory Safety Guidelines:
1. No eating or drinking in the laboratory at any time.
2. Playing or "horse play" in the laboratory is forbidden.
3. Read all signs and labels carefully.
4. Use personal protective equipment (PPE) when told by your instructor.
5. Keep the work area clear of all materials except those needed for your work.
6. Be familiar with the location and proper use of fire extinguishers, fire blankets, first aid kits, spill kits, etc. in the laboratory.
7. Make sure to read the assigned experiment before you start to work. Pay close attention to any cautions described in the laboratory application.
8. Students are responsible for the proper disposal of used material. Please make sure to place trash, broken glass, sharps, or any contaminated items in the appropriate containers.
9. Keep pathways clear by placing extra items (books, bags, etc.) on the shelves or under the laboratory tables. If under the tables, make sure that these items cannot be stepped on.
10. Application of cosmetics are prohibited in the laboratory.
11. Never do unauthorized experiments.
12. Keep sinks free of paper or any debris that could interfere with drainage.
13. Do not place fingers or objects, such as pencils, labels, tape, etc., in your mouth when working in the laboratory.
14. Clean up your laboratory work area before leaving.
15. Wash hands with soap and water before leaving the lab and before eating.

16. Follow your instructor's guidelines for using all lab equipment.
17. Notify your instructor if you have any allergies or medical condition that may require special precautionary measures while in the laboratory.
18. In the case of spills: Report at once to your instructor. Cover spilled material with a paper towel and soak with disinfectant. Leave for 20 minutes. Then discard the material in the biohazard buckets.

Accidents and Injuries:
1. Learn the location of the fire extinguisher, eyewash station, first aid kit and safety shower.
2. Report all accidents (spill, breakage, etc.) or injuries (cut, burn, etc.) to the instructor immediately, no matter how trivial it may seem. Do not panic!
3. Long hair (chin-length or longer), dangling jewelry, are hazards in the laboratory. Therefore, long hair must be tied back, and dangling jewelry must be secured. Long hair must be tied back to prevent it from catching fire.
4. If you or your laboratory partner is hurt, immediately (and loudly) call the instructor. Please do not panic!
5. Keep your lab space clean and organized to prevent accidents or injuries.
6. Do not lean, hang over or sit on the laboratory tables.
7. Be careful when lifting heavy objects.
8. Never work alone in the laboratory, unless stated otherwise by your instructor.
9. Do not leave an on-going experiment unattended.
10. Maintain unobstructed access to all exits, fire extinguishers, electrical panels, gas shout offs, emergency showers, and eyewash stations.
11. Do not pour any hazardous material down the sink.
12. Non-disposable contaminated materials are often autoclaved and reused.
13. If someone vomits or cuts him or herself you are to supply only "indirect assistance".
14. During outdoor activities be aware of any dangers in the area, venomous animals, poisonous plants, etc.

Clothing:
1. Any time chemicals, heat, or glassware are used, students will wear safety goggles.
2. Wear safety glasses or face shields when working with hazardous materials, equipment, and/or dissecting specimens.
3. Contact lenses may not be worn in the laboratory.
4. Loose or baggy clothing should be secured so they do not get caught in a flame or chemicals.
5. Lab coats or aprons should be worn during laboratory experiments, unless otherwise stated by your instructor.
6. Shorts and sandals should not be worn in the laboratory at any time, unless otherwise stated by your instructor.
7. Wear gloves when using any hazardous or toxic agent.

Chemical Safety:
1. Treat every chemical as if it were dangerous or hazardous.
2. Make sure all chemicals are clearly and currently labeled.
3. Avoid handling chemicals with fingers.
4. If a chemical should splash in your eye(s) or on your skin, immediately flush with running water for at least 20 minutes. Immediately (and loudly) yell out the instructor's name to get his or her attention.
5. Never remove chemicals or other materials from the laboratory, unless otherwise stated by your instructor.
6. Check the label on all chemical bottles twice before removing any of the contents.
7. Never return unused chemicals to their original container. (Try for the correct amount and share any excess.)
8. Use volatile and flammable compounds only in a fume hood.
9. Never allow a solvent to come in contact with your skin. Always use gloves.

10. Do not taste, or smell any chemicals, especially a solvent. Read the label on the solvent bottle to identify its contents.
11. Dispose of chemical waste according to your instructor's directions.
12. Clean up spills immediately.
13. Know the location of the Material Safety Data Sheet (MSDS) in your laboratory

Handling Glassware and Equipment:
1. Never handle broken glass with your bare hands
2. Use a brush and dustpan to clean up broken glass.
3. Examine glassware before each use.
4. Never use chipped, cracked, or dirty glassware. Check your glassware for cracks and chips each time you use it. Cracks could cause the glassware to fail during use and cause serious injury to you or your lab partners.
5. Do not immerse hot glassware in cold water. The glassware may shatter.
6. Report all breakage of glass or equipment to your instructor immediately.
7. If a piece of equipment fails while being used, report it immediately to your instructor. Never try to fix the problem yourself because you could harm yourself and others.
8. Please make sure to place broken glass in designated container(s) for disposal.
9. Please place anatomical models back in their proper storage places once you are finished handling them, after each class period. Make sure that all moveable or detachable parts are place back in their proper positions.
10. Thoroughly clean all instruments that are exposed to human skin or body fluids.
11. Prepared microscope slides sure be cleaned and placed back in their proper slide set when students are finished handling.
12. Clean and cover microscopes, then place them back into their proper storage place. This ensures that it does its job, while extending its usefulness for as long as possible
13. If you do not understand how to use a piece of equipment, ask your instructor.

Dissecting a Specimen:
1. When making an observation, keep at least one foot away from the specimen.
2. Pointed dissection probes, scalpels, razor blades, scissors, and microtome knives must be used with great care, and placed in a safe position when not in use.
3. Conduct dissections in an appropriate physical environment with the proper ventilation, lighting, furniture, and equipment, including hot water and soap for cleanup. Washing hands is standard operating procedure at all times when a laboratory activity is completed. Without proper ventilation, students will be exposed to potential hazardous vapors from preservatives in specimens.
4. When dissecting smaller specimens, seal the bag after removing the specimen, so as to confine the preservative in the specimen bag.
5. Use personal protective equipment (PPE), such as gloves, chemical splash goggles, and aprons, all of which should be available. PPE is necessary and required when working with materials that can put eyes in harm's way; such as preservatives and body fluids.
6. Body parts or scraps of the specimen are not to be disposed of in the sink.
7. Containers designated for the disposal of sharps (scalpel blades, razor blades, needles; dissection pins, etc.) and containers designated for broken glass are present in each laboratory. Never dispose of any sharp object in the regular trash containers.
8. When cutting with a scalpel or other sharp instrument, forceps may be used to help hold the specimen. Never use fingers to hold a part of the specimen while cutting.
9. Scalpels and other sharp instruments are only to be used to make cuts in the specimen, never as a probe or a

pointer.
10. Dispose of dissecting pins or other sharp objects in the red sharps containers/bag, **NOT** in the regular trash.
11. Follow the directions of the instructor concerning the proper disposal of preserved specimens after they are finished being used.
12. Never ingest specimen parts
13. When dissecting, you should cut away from yourself.
14. Treat all living things with respect. Avoid causing unnecessary stress to living animals

Heating Substances:
1. Never use open flames in laboratory unless instructed by instructor.
2. Hair, clothing, and hands should be a safe distance from open flames or hot plates at all times.
3. Heated glassware remain very hot for a long time. They should be set aside in a designated place to cool, and picked up with caution. Allow plenty of time for hot apparatus to cool before touching it.
4. Anyone wearing acrylic nails should not be allowed to work with matches, lighted splints, or bunsen burners.
5. Never look into a container that is being heated.
6. Never point a test tube being heated at another student or yourself. Never look into a test tube while you are heating it
7. Do not place hot apparatus directly on the laboratory desk.
8. If you have a small fire in a container, (for instance, a small beaker full of alcohol has caught fire) find something you can use as a lid for the container. When the container is covered, the fire will quickly burn itself out.
9. If you have a small fire which is not in a container, move away from the fire and shout for help. You can use a fire extinguisher to put the fire out. If you ever need to use a fire extinguisher, remember the following (A) pull the pin, (B) aim to the side at first, (C) depress the handle, (D) sweep the spray from side to side across the BASE of the fire (where the fire meets the fuel), not just at the flames. When the fire is out, clean up the area.
10. Wear safety goggles to protect your eyes when heating substances.
11. If there is a large fire, shout for help and leave the area immediately. The fire alarm will probably sound. When it does, evacuate the building.
12. If your clothing is on fire. Please don't run. It will only fan the flames and make the fire worse. Instead, you should **STOP** moving, **DROP** to the ground (lie down), and **ROLL** on the ground to squash out the flames. Yell continuously. <u>Note:</u> If you want to help a person who is in this sort of trouble, don't use a fire extinguisher. You must never use a fire extinguisher on a human being. The chemicals in the extinguisher can be harmful.

LABORATORY SAFETY
LAB 1

CRASHCOURSE VIDEO(S):

Click on the video embedded within your online platform or enter the address below into your web browser:
1. https://youtu.be/VRWRmIEHr3A

(Please make sure to watch the video before continuing)

DEFINING KEY TERMS:

1. Binocular Microscope:
2. Biohazard:
3. Carcinogens:
4. Compound light microscope:
5. Condenser:
6. Contaminated:
7. Course Adjustment:
8. Debris:
9. Defibrillator or AED:
10. Diopter:
11. Dissecting Microscope:
12. Dissecting:
13. Empty magnification:
14. Fine Adjustment:
15. Fire Extinguisher:
16. First Aid Kit:

17. First Aid:

18. Flammable:

19. Focus Knob:

20. Hazard:

21. HAZMAT:

22. Illuminator:

23. Iris Diaphragm:

24. Irritant:

25. Mechanical Stage:

26. Methylene blue:

27. Microscope Arm:

28. Microscope Base:

29. Microscope Body Tube:

30. Microscope Coverslip:

31. Microscope Field of View:

32. Microscope Slide:

33. Monocular Microscope:

34. Mutagens:

35. Objective lens:

36. Ocular Lens/Eyepiece(s) :

37. Oil Immersion:

38. Optical lens wipe:

39. Oxidizing Substances:

40. Personal Protective equipment (PPE):

41. Prepared/permanent slide:

42. Radiation:

43. Radioactive Material:

44. Refraction:

45. Rotating Nosepiece:

46. Rotating Objective Turret:

47. Scanning Lens:

48. Sharps:

49. Specimen:

50. Squamous Epithelium cells:

51. Stage clip:

52. Stage Plate:

53. Stage:

54. Stereo Head:

55. Temporary/wet mount:

56. Teratogens:

57. Total Magnification:

58. Toxic:

59. Zooming Knob:

EXERCISE 1.1
IDENTIFYING LAB SAFETY HAZARDS

Purpose of exercise: To demonstrate proper behavior and safety concerns in the laboratory setting using scenarios.

Healthcare is involved, directly or indirectly, with the providing of health services to individuals. These services can occur in a variety of work settings, including hospitals, clinics, dental offices, out-patient surgery centers, birthing centers, emergency medical care, laboratories, home healthcare, and nursing homes. Healthcare professionals are exposed to several serious safety and health hazards on the job. They include bloodborne pathogens and biological hazards, potential chemical and drug exposures, needlestick injuries, waste anesthetic gas exposures, respiratory hazards, ergonomic hazards from lifting and repetitive tasks, laser hazards, hazards associated with laboratories, and radioactive material and x-ray hazards.

This is why it's important for Healthcare professionals to take the necessary precautions, document occurrences, and wear personal protective equipment (gowns, gloves, safety goggles, and face shields) to avoid contamination or injury.

Your experiences in this **laboratory** will prepare you for the safety hazards and mandates you may encounter in your prospective healthcare professions. Whether you're a medical assistant, pharmacy technician, clinical laboratory technician, surgical technician, radiological technician, or nurse, proper safety practices are vitally important.

Proper laboratory safety ensures the good reputation of the lab, healthcare facility, clinic, or pharmacy you work in. When people believe you work in a professional and safe environment, they're more likely to trust you what you do. And what you do for your individual healthcare employer also reflects well on the healthcare profession, as a whole.

DIRECTIONS: Using your handout on science safety rules, in each of the following situations, write "yes" if the proper safety procedures are being followed and "no" if they are not. Then give a reason for your answer.

1. Cayden cannot find matches to light his Bunsen burner. The student next to him picks up a lighted burner and says, "Here, you can use my flame to light your burner."

2. Mackenzie notices that the electrical cord on his microscope is frayed near the plug. She takes the microscope to her instructor and asks for permission to use another one.

3. The directions in lab manual instruct students to pour a small amount of hydrochloric acid into a beaker. Victoria puts on safety goggles and a lab apron before pouring the acid into the beaker.

4. Ronnie finds a paper clip on the floor, and becomes curious. He wants to know what will happen if he sticks the paper clip in the electrical outlet at his lab desk. His lab partners agree, and he performs his experiment.

5. While using ice in a lab, Michael puts a piece of ice down Victor's shirt. To get even, Travis grabs a handful and returns the favor.

DIRECTIONS: Look at each of the following drawings and **explain** what rules the individuals are following what rules they are not following correctly (unsafe).

6.

7.

8.  9.

10.  11.

DIRECTIONS: Read each of the statements below. If the statement is true, then write "True" in the space provided. If the statement is false, then write "False" and fix the statement so that it is true.

12. _____ While in the lab, it is okay to begin touching or using the items on the lab tables without being instructed to do so, as long as you know how to use and handle the equipment.

13. _____ If something is spilled or broken in the lab, just cover it up or kick it out of the way if no one saw you do it.

14. _____ If an acid gets on your skin, the best thing to do is to tell your instructor immediately, and flush the area with lots of water.

15. _____ When dissecting, you should cut away from yourself.

16. _____ The best place to throw away garbage or unwanted items is in the lab sink, because instructors love to pick up other people's messes.

17. _____ Long hair should be tied back during labs so it does not get in the way.

DIRECTIONS: Read each of the statements below and write the correct responds.

18. According to the dress code, what should you wear in the laboratory?
19. When should you wear goggles?
20. In what cases might you be asked to leave the laboratory?
21. Can you eat in the laboratory? Explain your answer.
22. What should you do if:
    a. You broke a beaker and cut your finger.

b. Chemicals have splashed on your face.
c. The fire alarm sounds.
d. Your lab manual has caught on fire.
e. Your shirt has caught on fire.
f. Chemicals have spilled on your pants.

23. When is it permissible to eat or drink in the laboratory?

24. Is the following statement true? Explain your answer. If the instructor is eating or drinking in the laboratory, then the students have implicit permission do so as well.

25. If someone vomits or cuts him or herself you are to supply only indirect assistance. What does "indirect assistance" mean? Give examples.

26. What kinds of things should not be poured down the sink? Give examples.

DIRECTIONS: After reading each of the following rules, give a reason as to why that rule is should be followed?

	Rule	What could happen if this rule is not followed properly?
27.	At the beginning of most laboratories, your instructor will engage in a pre-lab discussion. Many safety procedures will be discussed during these discussions.	
28.	Keep all books, papers, and other flammable materials away from open flames or dangerous chemicals.	
29.	Tie back long hair when you are working with an open flame. Pipe cleaners, rubber bands, and string are useful for this purpose.	
30.	Never use chemicals from an unlabeled container. Do not taste, smell, or touch chemicals unless specifically instructed by your instructor to do so.	
31.	Wear safety goggles during experiments involving heating or hammering or while using acids or bases. If you do not have goggles on, stay away from students that are experimenting.	
32.	Point the open end of a test tube or flask away from yourself and others while heating it. Never heat a closed container.	

33.	Use squeeze bottles and droppers only for their intended purpose.	
34.	Another common accident is picking up red hot materials. Take proper precautions against this.	
35.	No material should be left in the sinks; i.e. paper, beakers, etc.	
36.	Never place pencils, pens, or other materials in your mouth.	
37.	Never return excess chemicals back to their container.	
38.	Keep volatile liquids and reagents away from the Bunsen Burner flame or other heating source.	
39.	Know where all laboratory safety equipment is located in case you need it.	
40.	Most chemical spills are best handled by washing the affected area with water as quickly as possible. Call your instructor for assistance if necessary. Severe spills may require the removal of clothing.	
41.	Put out any fires immediately. Call your instructor for assistance if necessary.	
42.	In an emergency situation an all too common response is panic. If you observe another student in trouble, tell them what to do, and assist them in doing it.	
43.	You are responsible for keeping your laboratory area and completely neat and clean.	
44.	When dissecting, you should cut away from yourself.	

EXERCISE 1.2
SIGNS & SYMBOL SAFETY

Purpose of exercise: To recognize laboratory safety symbols used in the text, experiments, and in the classroom.

One of the more remarkable achievements in the progression of humankind is how we learned to communicate through symbols. In our language, images, and gestures we convey messages that must be learned and interpreted by others. A symbol is something that represents or stands in for something else. The term symbol originates from the Greek word "symbolon" meaning token or sign. Symbols must be learned as they represent, stand in for, or suggest something else such as an idea or object.

Have you ever come across a sign or symbol in a healthcare facility, but did not know its meaning? If so, you're not alone. Signs and symbols are supposed to clearly convey important information about the hospital setting. When a hospital needs to inform visitors, employees and patients on what to do in case of emergencies, where rooms are located or how to find a particular department, they use signs or symbols to convey this information. In order to be effective, these signs or symbols must draw attention and communicate clearly. Hospitals are usually very large facilities and without signs, visitors would have a difficult time figuring out where things were located.

Similar to healthcare facilities, laboratories are places where dangerous events can occur. Many people are unaware of the dangers lurking in a laboratory. To alert you of hazards, there are dozens of signs and symbols placed throughout the laboratory to guide you safely. To help you get aware of the various meanings associated with each sign or symbol, this guide has been compiled. Please complete the exercise below, to familiarize yourself with these signs and symbols.

Meaning	Symbol	Meaning	Symbol	Meaning	Symbol
Indicates eyewash station		Gas under pressure		Toxic substances	
Indicates safety shower		Indicates fire extinguisher		Indicates carcinogens, mutagens, teratogens, respiratory sensitizers and substances with target organ toxicity	
Indicates first aid station		Indicates fire hose		Indicates irritant or the general symbol for a potentially harmful chemical	

Indicates defibrillator or AED		Indicates radioactive material		Indicates flammable substances	
Indicates flammable gas. HAZMAT Class 2.1		Indicates biohazard		Indicates explosives or an explosion hazard	
Indicates fire blanket		Universal recycling symbol		Indicates corrosive materials	
Prohibition		Indicates oxidizing substances		Indicates environmental hazard	
Indicates respiratory protection is required		Indicates mandatory eye or face protection		Indicates mandatory use of protective footwear	
Wear gloves or other hand protection		Indicates mandatory use of protective clothing		Caution, no food or drink in area	
Indicates ear protection is required		Indicates the presence of a low temperature or cryogenic hazard		Indicates a hot surface	
Indicates Non-ionizing radiation		Warning symbol		Indicates the presence of an optical radiation hazard	

Indicates danger	DANGER	Warns of laser radiation	LASER RADIATION	Indicates the presence of a high voltage hazard	
Indicates caution	CAUTION	High level source of Radiation		No smoking	
Wheelchair access		The four divisions are color-coded with red indicating flammability, blue indicating level of health hazard, yellow for chemical reactivity, and white containing codes for special hazards. Each health, flammability and reactivity is rated on a scale from 0 (no hazard) to 4 (severe risk)			

EXERCISE 1.3
IDENTIFYING LABORATORY EQUIPMENT

Purpose of exercise: To properly identify equipment and there usage in the laboratory setting.

Click on *Exercise 1.3* within your online platform or enter the address below into your web browser:
http://www.sciencegeek.net/Chemistry/taters/labequipment.htm

Complete the activity provide by the website link. Once you are done identifying the laboratory equipment in the interactive exercise, please move on and complete the remaining questions located in your laboratory manual for this exercise.

1. Name some laboratory tools that are used to make measurements.

2. How might glassware be used differently?

3. When might you need to use a thermometer in the lab?

4. Which laboratory tools can be used to magnify small objects so they can be seen more easily?

5. Which laboratory tools are useful when dissecting a specimen?

6. What laboratory tool or tools would you use to measure?

7. How do laboratory tools improve the observations made by a scientist?

EXERCISE 1.4
MICROSCOPY

Purpose of exercise: To familiarize students with the care, parts, function, and use of the Compound Light Microscope and Stereoscopic Dissecting Microscope.

Good microscopy involves three main factors: resolution, magnification, and contrast. The resolution, or resolving power of a microscope, which will increase when the wavelength of the light source is reduced. A microscope's total magnification is a combination of the ocular lens and the objective lens. "Empty magnification" occurs when the image continues to be enlarged, but no additional detail is resolved. Different microscopes will come with a range of magnifications, such as light microscope ranging from 40 to 1000 and the stereomicroscope from 7 to 30. Staining with suitable stains will increase the contrast of a specimen to be observed under a microscope.

There are two main types of light microscope, compound and stereoscopic dissecting microscope. While the compound microscope is ideal for viewing thin slices of tissues that can be mounted on a slide, there are biological specimens that are too large and/or thick to be mounted on a slide and viewed with the compound microscope (yet too small for the naked eye). In this case, you will need to use the stereoscopic dissecting microscope or "dissecting microscope" for short. Two advantages of this microscope are

1) you can manipulate your specimen (turn, flip, dissect) using your hands or tools while viewing it under magnification (hence term "dissecting"), and
2) by looking through both oculars you can see the image in three dimensions ("stereoscopic").

While the total magnifications possible on this microscope are low, they provide the advantages of a very large field of view and a very thick depth of focus. This will allow you to see most, if not all, of your specimen clearly and in three dimensions.

Light Microscope Safety, Maintenance, and Usage

Microscope Safety
1. Never disassemble the microscope as doing so may cause electric shock or damage to the microscope
2. Allow the halogen bulbs to cool before touching. Halogen bulbs become extremely hot and may cause burns if touched
3. To avoid electric shock or damage to the instrument, unplug the microscope before replacing the bulb.
4. Use only the prescribed halogen or fluorescent bulb
5. Turn off and unplug the microscope before moving.

Carrying the Microscope
1. Always lift the microscope with two hands: one hand on the arm, the other hand supporting the base.

Cleaning the Microscope
1. Dust should be cleaned off with a soft brush
2. Clean smudges, fingerprints, and oils from the lens with clean lens paper or a soft clean cloth moistened with a small amount of isopropyl alcohol.

3. Clean the microscope body and stand using a moist, soft cloth with a small amount of detergent.
4. Use water only on plastic surfaces
5. Do not use paint thinner or other solvents
6. Do a final wipe with a moist soft cloth
7. Dry all surfaces after cleaning

Storage
1. Always cover the microscope with a dust cover/jacket when not in use
2. Store in a dry place
3. In humid or moist environments, it is wise to store the microscope in a waterproof container with a drying agent
4. Do not touch the optical lens with bare hands or fingers
5. Do not store the microscope in direct sunlight.
6. Store microscope set on the lowest objective with the nosepiece turned down to its lowest position (use the coarse adjustment knob)

Usage
1. Set your microscope on a tabletop or other flat, sturdy surface where you will have plenty of room to work.
2. Plug the microscope's power cord into an outlet.
3. Switch on your microscope's light source and then adjust the diaphragm to the largest hole diameter, allowing the greatest amount of light through. If you have an iris diaphragm, slide the lever till the most light comes through.
4. Rotate the nosepiece to the lowest-power objective, typically the scan objective lens. *(It has a red-band).*
5. Place a microscope slide on the stage, either under the stage clips or clipped onto the mechanical stage.
6. Adjust the large coarse focus knob until the specimen is in focus. Slowly move the slide to center the specimen under the lens, if necessary. If you have a mechanical stage, do this by gently turning the slide control knobs.
7. Adjust the small fine focus knob until the specimen is clearly in focus.
8. Adjust the diaphragm to get the best lighting. Start with the most light and gradually lessen it until the specimen image has clear, sharp contrast.
9. Scan the slide (right to left and top to bottom) at low power to get an overview of the specimen. Then center the part of the specimen you want to view at higher power.
10. Once you have the image in view, rotate the nosepiece to view it under different powers. Rotate the nosepiece to the 10x objective for 100x magnification. Refocus and view your specimen carefully. Adjust the lighting again until the image is most clear (you will need more light for higher power). Repeat with the 40x objective for 400x magnification, then 100x objective for 1000x magnification. (To calculate the power of magnification, multiply the power of the ocular lens by the power of the objective.)
11. If your microscope has a 100x oil-immersion lens, you'll need to put 1-2 drops of immersion oil over the slide coverslip (the piece of glass over the middle of the slide) before viewing it at highest power. We can see better details with higher the powers of magnification, but we cannot see as much of the image.
12. Move the 100x objective lens into position, and then slowly move the stage up until the lens makes contact with the oil. Continue focusing with the coarse focus knob until the color or blurred outline of the specimen appears. Finish focusing with the fine focus knob. With the 100x lens, you will be able to see additional cell detail, but you will need to take extra care with focus and contrast for a clear image. When you are done using the slide, clean the oil off of the slide and the lens with lens cleaning paper and solution.

In this exercise you will learn how to properly identify parts and the use of microscopes.

1. Click on *Exercise 1.4 (a)* within your online platform or enter the address below into your web browser:
   https://www1.udel.edu/biology/ketcham/microscope/testFLV8.html

2. After you have watched the video in the link above. Click on *Exercise 1.4 (b)* within your online platform or enter the address below into your web browser:
   http://www1.udel.edu/biology/ketcham/microscope/scope.html

   Familiarize yourself with the virtual microscope at this link. It has all the same controls found on the real thing. Virtual Microscope controls:
   1. turn knobs (click and hold on upper or lower portion of knob)
   2. throw switches (click and drag)
   3. turn dials (click and drag)
   4. move levers (click and drag)
   5. changes lenses (click and drag on objective housing)
   6. select a specimen (click on a slide)
   7. adjust oculars (in "through" view, with the light on, click and drag to move oculars closer or further apart)

3. Now let's work with the microscope a little more. Click on *Exercise 1.4 (c)* within your online platform or enter the address below into your web browser:
   https://www.brainpop.com/games/virtuallabsusingthemicroscope/

   Once you are comfortable with your understanding of the microscope, please label the parts of the Compound Light Microscope in your laboratory manual.

Label the parts of the Compound Light Microscope

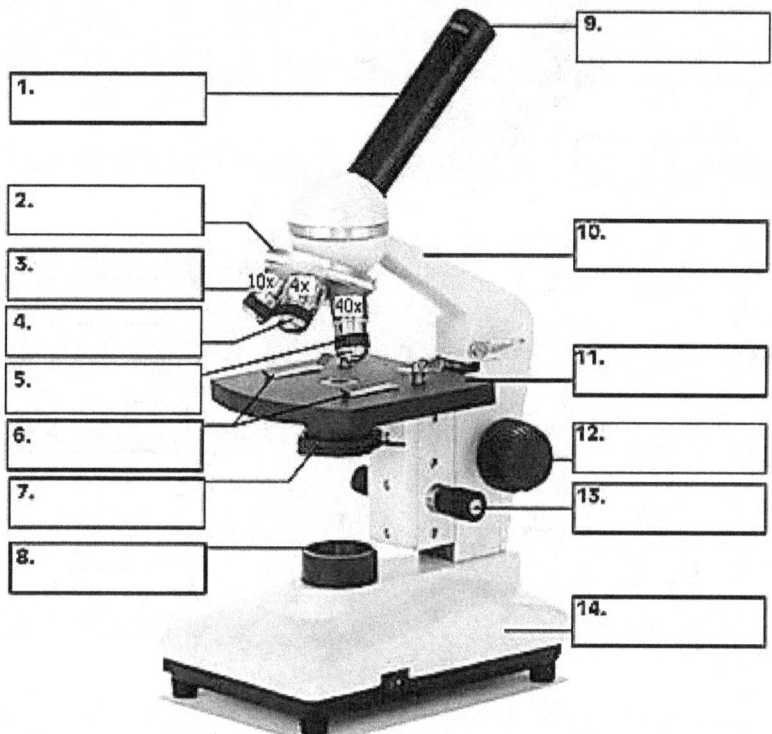

4. Click on *Exercise 1.4 (d)* within your online platform or enter the address below into your web browser: **http://micro.magnet.fsu.edu/primer/flash/nikonsmz1500/index.html**

Familiarize yourself with the virtual Stereoscopic Dissecting Microscope at this link. Choose and view any of the objects in the drop down list. Once you are comfortable with your understanding of the microscope, please label the parts of the Stereoscopic Dissecting Microscope in your laboratory manual.

Label the parts of the Stereoscopic Dissecting Microscope

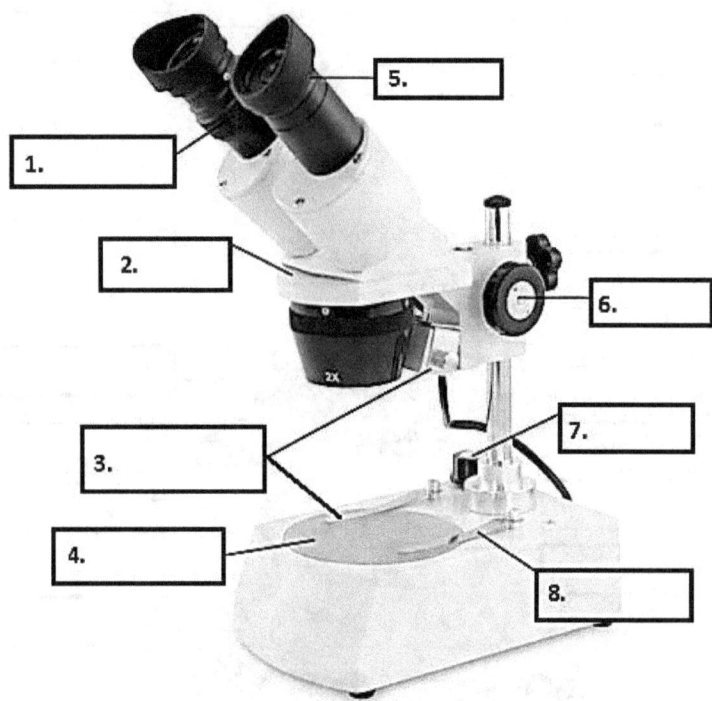

5. In your laboratory manual, calculate the magnifications for each missing item in the table below.

Eyepiece Magnification	Objective Magnification	Total Magnification
10x	40x	
10x		900x
	20x	1200x

6. What happens to our view of an image as you increase the power of magnification?
7. When focusing a specimen, you should always start with the _____ objective.
8. When using the high power objective, only the _____ knob should be used.
9. The two types of microscopes used in most biology laboratories are the _____ & _____.
10. You should carry the microscope by the _____ and the _____.
11. The objectives are attached to what part of the compound microscope (it can be rotated to click lenses into place?)
12. A microscope has an ocular objective of 10x and a high power objective of 60x, what is the microscope's total magnification?
13. Explain the steps used to create a wet mount.

EXERCISE 1.5
MICROSCOPY

Purpose of exercise: To become familiar with the operation and limitations of light microscopes while examining a prepare specimen.

In this exercise, you will view a smear of the cheek cells. The cells on the inside of your cheek are called squamous epithelium cells and can be easily viewed with a Compound Light Microscope and Stereoscopic Dissecting Microscope.

Click on *Exercise 1.5* within your online platform or enter the address below into your web browser:
http://www1.udel.edu/biology/ketcham/microscope/scope.html

* Please choose the cheek smear slide from the choices given, to observe under the virtual compound light microscope.

Due to this being an online class, you are not required to physically make the smear of the cheek cells. However is in important that you know the process. Therefore, read the information below to further your understanding of how to prepare a microscope slide for viewing.

How to make the microscope slide:
To make a cheek smear, take a clean toothpick and gently scrape the inside of your cheek. Then wipe that part of the toothpick in the center of your slide. Hold the coverslip with one end flush on the slide and gently wipe the edge of the coverslip along the middle of the slide's surface. This will smear the cells along the slide, making a layer thin enough to view clearly. Let the smear air dry.

Once your smear is dry, add a drop of methylene blue stain to the center of the smear so you will be able to see the cells more clearly. Gently set a coverslip over the smear and scan your slide under low power to locate the cells, then observe them more closely under high power.

EXERCISE 1.6
MICROSCOPY

Purpose of exercise: To observe, draw, and describe prepared specimens under the Compound Light Microscope and Stereoscopic Dissecting Microscope.

1. Click on *Exercise 1.6 (a)* within your online platform or enter the address below into your web browser:
   http://www1.udel.edu/biology/ketcham/microscope/scope.html

Please choose two of the specimen slides from the choices given to observe under the Virtual Compound Light Microscope. View your specimens under 4x (scan), 10x (low power), and 40x (high power) magnifications.

Name of specimen one: _____

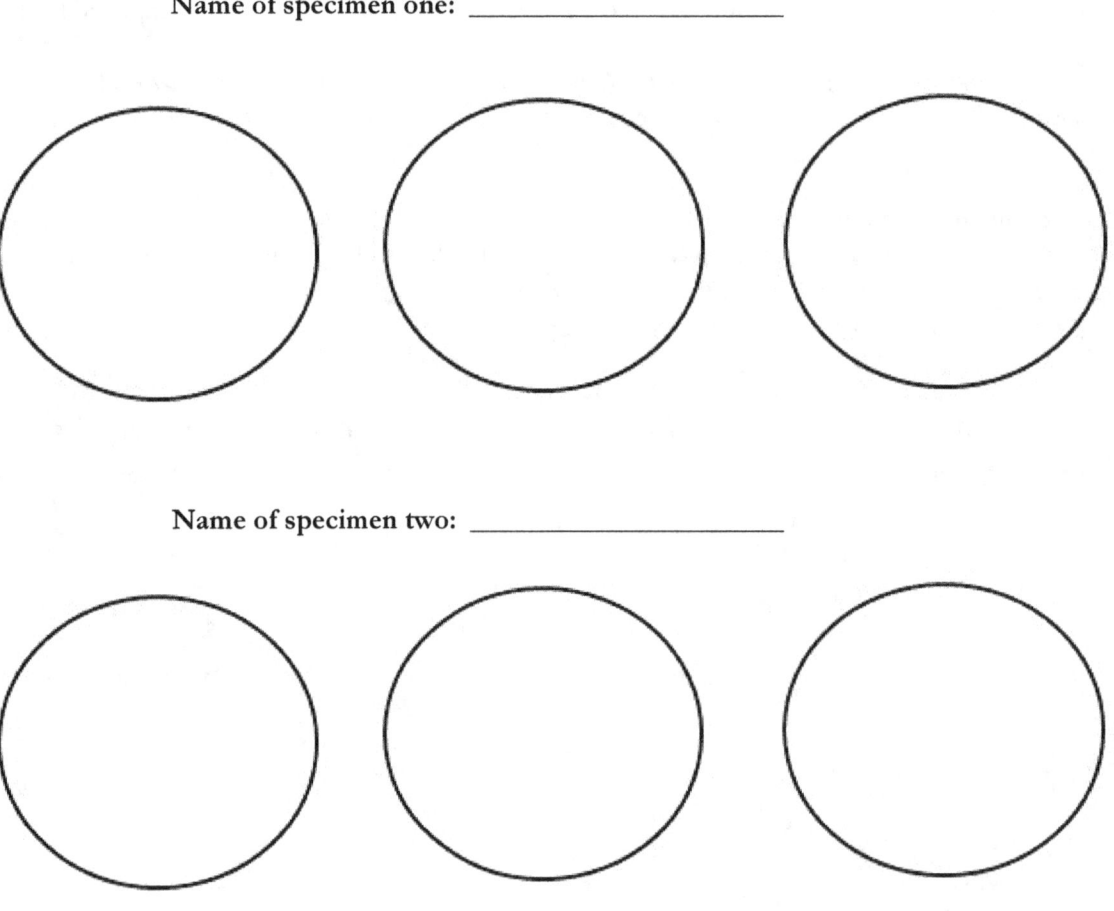

Name of specimen two: _____

2. Click on *Exercise 1.6 (b)* within your online platform or enter the address below into your web browser:
   http://micro.magnet.fsu.edu/primer/flash/nikonsmz1500/index.html

Please choose the Flower Bud from the specimen choices given to observe under the Virtual Dissecting Microscope. Place your setting to the following: **Zoom Mode:** Smooth, **Eyepieces:** 10x, **Objective:** 1x, **Field of View:** 29.30mm,

Working Distance: 54mm, **Magnification:** 7.50x, **Focus and Intensity:** Both should be equal (halfway). Please draw what you see.

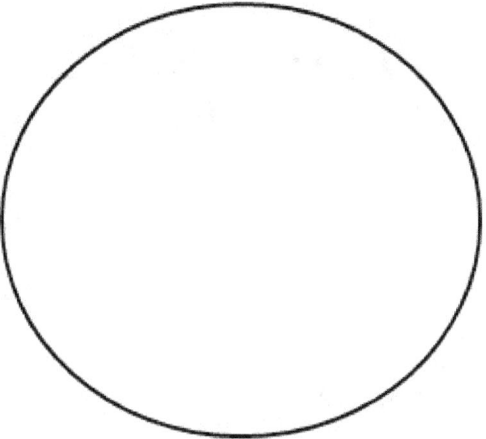

BODY ORGANIZATION
LAB 2

CRASHCOURSE VIDEO(S)

Click on the video embedded within your online platform or enter the address below into your web browser:
1. https://youtu.be/uBGl2BujkPQ

(Please make sure to watch the video before continuing)

DEFINING KEY TERMS:

The healthcare industry has its own terminology. The purpose of this language is not to confuse, but rather to increase precision and reduce medical errors. Using standard anatomical terminology relative to body organization ensures that healthcare providers have a common method of communicating and avoids confusion when identifying body parts, locations, and positions correctly. An understanding of anatomical terminology and body organization is a requirement for all healthcare providers.

The top <u>nine</u> terms have been paired and identified with its opposite.			
1. Superior/Inferior	2. Anterior/Posterior	3. Superficial/Deep	4. Medial/Lateral
5. Ventral/Dorsal	6. Proximal/Distal	7. Right/Left	8. Bilateral/Ipsilateral /Contralateral
9. Cephalad, Cranial/ Caudal	10. Axial / Appendicular		

Body sections and planes
Medical imaging techniques such as sonography, CT scans, MRI scans, or PET scans are one of the primary applications of body planes. By imaging a patient in standard anatomical position, a radiologist can build an X-Y-Z axis around the patient to apply body planes to the images. The planes can then be used to identify and locate the positions of the patient's internal organs. Individual organs can also be divided by planes to help identify smaller structures within that organ. Anatomical change during embryological development is also described and measured with body planes.

1. Frontal or Coronal	2. Transverse or Horizontal	3. Median/Midsagittal	4. Parasagittal plane
5. Oblique sections	6. Sagittal plane	7. Longitudinal plane	

The anatomical reference point is a standard body position called the "**anatomical position**". <u>**In the anatomical position, the body is erect, the palms of the hand face forward, the thumbs point away from the body, and the feet are slightly apart.**</u>

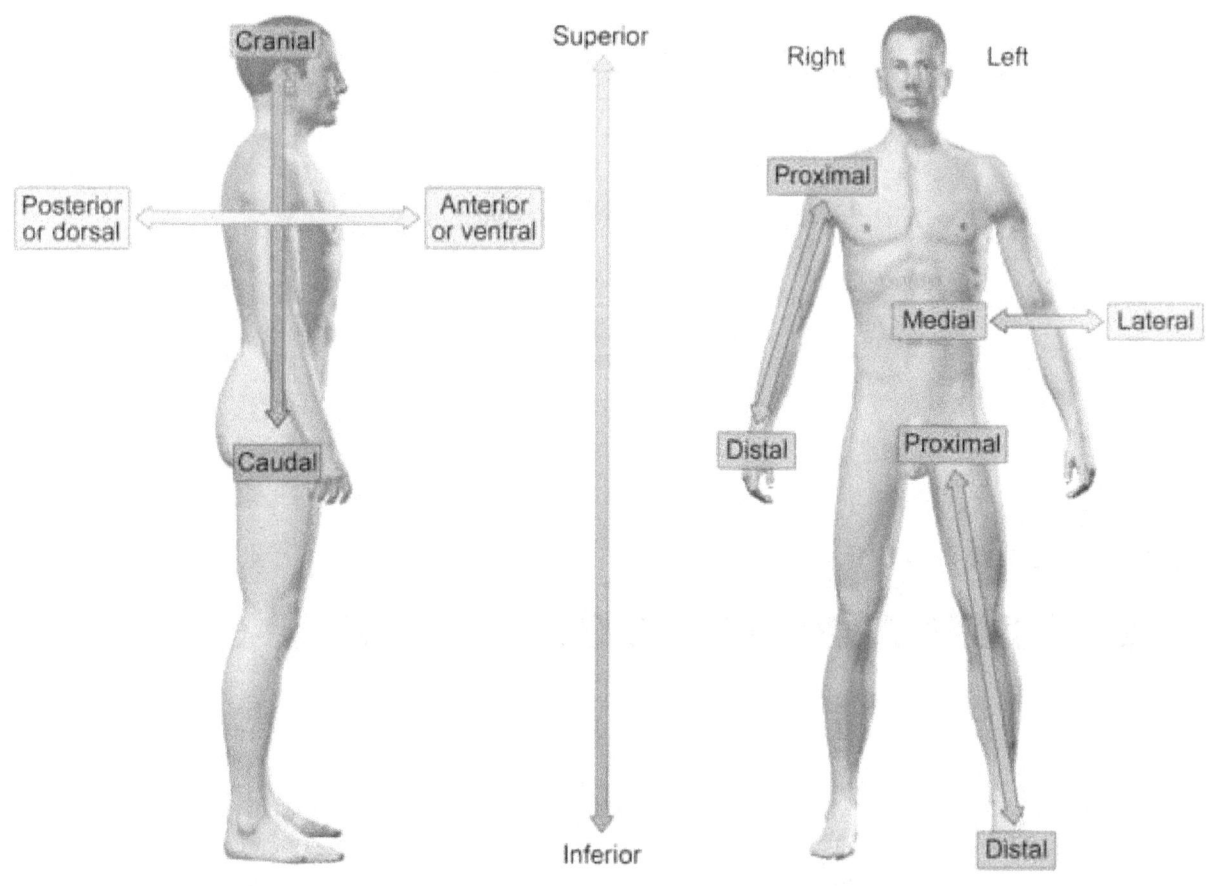

Lateral view   Anterior view

Many vital organs are suspended in internal chambers called body cavities. A cavity, is a hollow space within the body. The **dorsal cavity** and the ventral cavity are the two largest body cavities. The dorsal cavity is located at the posterior aspect of the body. Posterior means that it runs along the back of the body. The **ventral cavity runs** along the anterior aspect of the body. The figure below illustrates the **major body cavities** and their subdivisions.

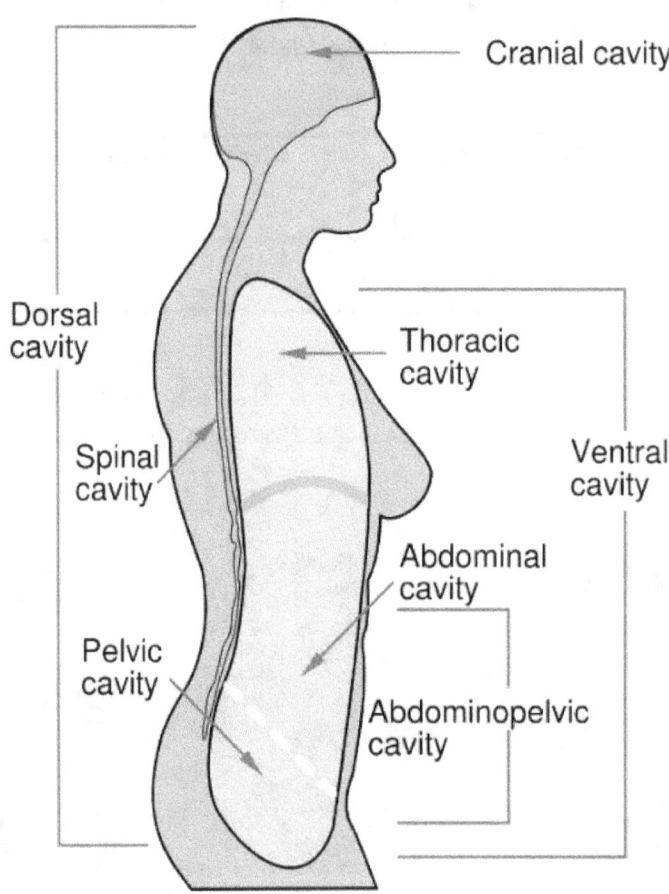

EXERCISE 2.1
CHARACTERISTICS OF LIFE

Purpose of exercise: List and describe the major characteristics of life.

Is Morris Alive? Use the characteristics of life discussed in class to answer the questions throughout this story.

1. Morris was a normal, healthy man. There was nothing in his life to indicate that he was anything different from anyone else. When he completed high school, he obtained a job in a factory, operating a machine press. On this job he had an accident and lost his hand. It was replaced with an artificial hand that looked and operated almost like a real one.
Is Morris Alive? Explain your answer.

2. Soon afterward, Morris developed a severe intestinal difficulty, and a large portion of his lower intestine had to be removed. It was replaced with an elastic silicon tube.
   Is Morris Alive? Explain your answer.

3. Everything looked good for Morris until he was involved in a serious car accident. Both of his legs and his good arm were crushed and had to be amputated. He also lost an ear. Artificial legs enabled Morris to walk again, and an artificial arm replaced the real arm. Plastic surgery enabled doctors to rebuild the ear.
   Is Morris Alive? Explain your answer.

4. Over the next several years, Morris was plagued with internal disorders. First, he had to have an operation to remove his aorta and replace it with a synthetic vessel. Next, he developed a kidney malfunction, and the only way he could survive was to use a kidney dialysis machine (no donor was found for a kidney transplant). Later, his digestive system became cancerous and was removed. He received nourishment intravenously. Finally, his heart failed. Luckily for Morris, a donor heart was available, and he had a heart transplant.
   Is Morris Alive? Explain your answer.

5. It was now obvious that Morris had become a medical phenomenon. He had artificial limbs, nourishment was supplied to him through his veins; therefore, he had no solid wastes. The kidney dialysis machine removed all waste material. The heart that pumped his blood to carry oxygen and food to his cells was not his original heart. But Morris's transplanted heart began to fail. He was immediately placed on a heart-lung machine. This supplied oxygen and removed carbon dioxide from his blood, and it circulated blood through his body.
   Is Morris Alive? Explain your answer.

6. The doctors consulted bioengineers about Morris. Because almost all of his life-sustaining functions were being carried on by machine, it might be possible to compress all of these machines into one mobile unit, which would be controlled by electrical impulses from Morris's brain. This unit would be equipped with mechanical arms to enable him to perform manipulative tasks. A mechanism to create a flow of air over his vocal cords might enable him to speak. To do all this, they would have to amputate at the neck and attach his head to the machine, which would then supply all nutrients to his brain. Morris consented, and the operation was successfully performed.
   Is Morris Alive? Explain your answer.

7. Morris functioned well for a few years. However, a slow deterioration of his brain cells was observed and was diagnosed as terminal. So the medical team that had developed around Morris began to program his brain. A miniature computer was developed: it could be housed in a machine that was humanlike in appearance,

movement, and mannerisms. As the computer was installed, Morris's brain cells completely deteriorated. Morris was once again able to leave the hospital with complete assurance that he would not return with biological illness.

Is Morris Alive? Explain your answer. If Morris is not alive at the story's end, exactly when did Morris stop being alive?

EXERCISE 2.2
ANATOMICAL PARTS

Purpose of exercise: To properly use the terms that describe relative positions, body sections, and body regions.

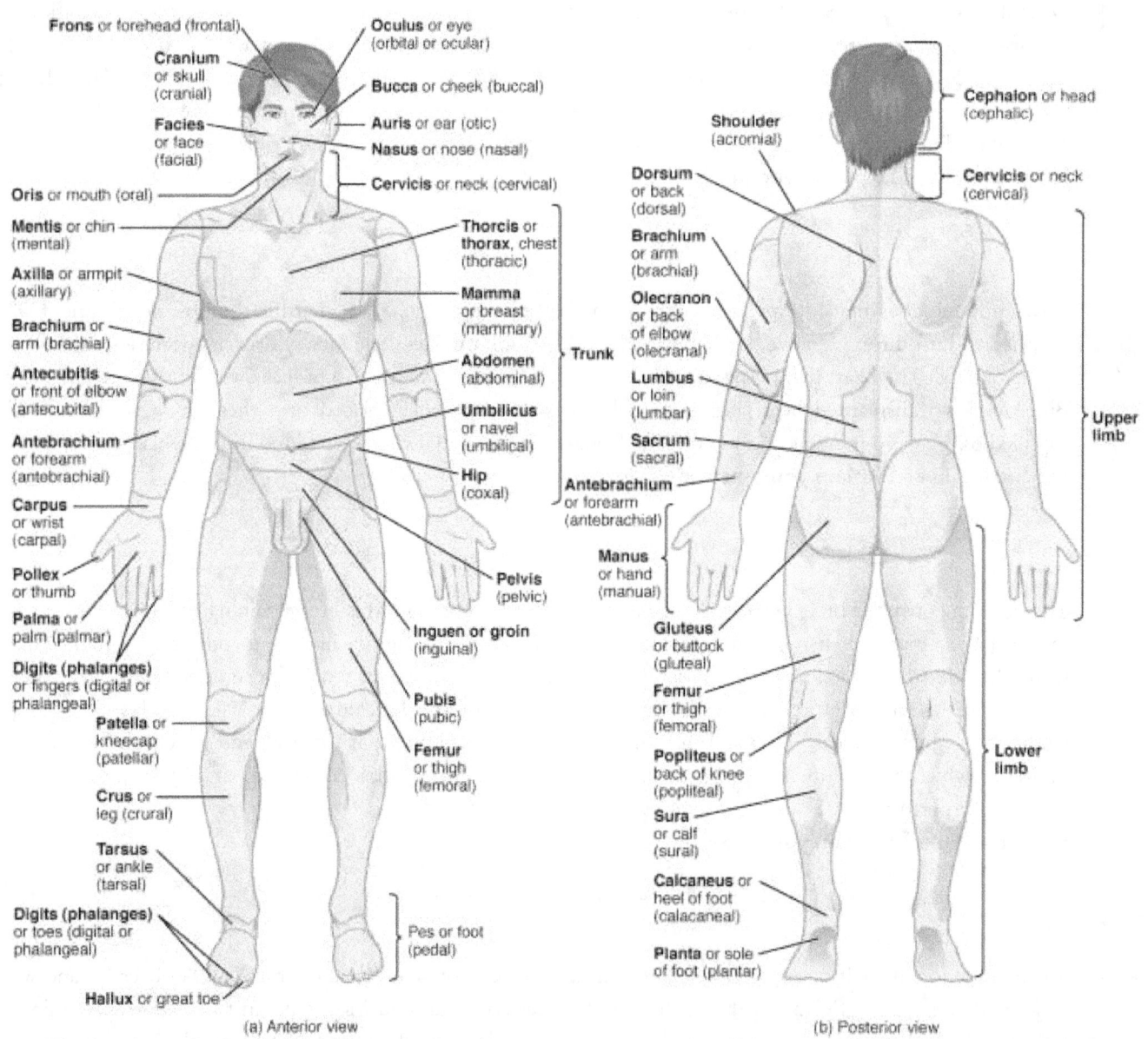

(a) Anterior view   (b) Posterior view

Complete the table in the spaces provided below using the image in this excise as a reference.

Surface Anatomy (Landmarks)	
Anterior Landmarks:	
Above the Waist 1. Abdominal:_____ 2. Acromial:_____ 3. Antebrachial:_____ 4. Antecubital:_____ 5. Axillary:_____ 6. Brachial:_____ 7. Buccal:_____ 8. Carpal:_____ 9. Cephalic:_____ 10. Cervical:_____ 11. Digital:_____ 12. Frontal:_____ 13. Mammary:_____ 14. Mental:_____ 15. Metacarpal:_____ 16. Nasal:_____ 17. Oral:_____ 18. Orbital:_____ 19. Otic:_____ 20. Palmar:_____ 21. Pectoral:_____ 22. Sternal:_____ 23. Umbilical:_____	**Below the Waist** 1. Coxal:_____ 2. Digital:_____ 3. Femoral:_____ 4. Genital:_____ 5. Inguinal:_____ 6. Patellar:_____ 7. Pedal:_____ 8. Sural:_____ 9. Tarsal:_____
Posterior Landmarks	
Above the Waist 1. Acromial:_____ 2. Cubital:_____ 3. Dorsum:_____ 4. Lumbar:_____ 5. Occipital:_____ 6. Vertebral:_____	**Below the Waist** 1. Calcaneal:_____ 2. Gluteal:_____ 3. Perineal:_____ 4. Plantar:_____ 5. Popliteal:_____ 6. Sacral:_____

EXERCISE 2.3
ANATOMICAL PARTS

Purpose of exercise: To name the major cavities, organ systems, and list the organs associated with each.

Click on *Exercise 2.3* within your online platform or enter the address below into your web browser: https://www.biodigital.com/

- *(This is a free site, but you will need to sign up with your name and a validated email address. You can use your personal email address or school email address. MAKE SURE TO CREATE A PASSWORD YOU CAN REMEMBER. I WOULD SUGGEST WRITING IT DOWN AND KEEPING IT IN A SAFE PLACE. YOU WILL USE IT AGAIN IN PART II OF THIS COURSE)*

Click **SIGN UP**. Once you've provided the appropriate information, you are now able to access BioDigital's website content. Click **LOG IN** using the email and password you created. Then click **SIGN IN**. On the left of your screen choose from the systems listed. Using your lab manual as a reference, place a check next to the box once you've identified the parts:

(Your instructor may choose to upload pictures of the Dissectible Human Torso Model in your school's online platform. If this is the case, you can still utilize the listed structures below as a reference to identify.)

☐ skin ☐ brain ☐ teeth ☐ tongue ☐ larynx ☐ thyroid gland ☐ trachea ☐ bronchi ☐ lungs ☐ liver ☐ stomach ☐ large intestine ☐ small intestine ☐ larynx ☐ thyroid gland ☐ superior vena cava ☐ inferior vena cava ☐ aortic arch ☐ abdominal aorta ☐ heart ☐ lungs ☐ esophagus ☐ diaphragm ☐ adrenal glands ☐ kidneys ☐ stomach ☐ spleen ☐ lymph node ☐ lymphatic vessels ☐ urinary bladder ☐ ureter ☐ pancreas

EXERCISE 2.4
ANATOMICAL PARTS

Purpose of exercise: To name the major cavities, organ systems, and list the organs associated with each.

Using nine terms (*From the Defining Key Terms section*), write twelve statements for each of the six pairs of directional terms. The statements should show the relationship of each pair.

Example: *"The wrist is **distal** to the elbow, but the shoulder is **proximal** to the elbow."*

EXERCISE 2.5
ANATOMICAL PARTS

Purpose of exercise: To properly use the terms that describe relative positions, body sections, and body regions.

1. Using the key bold italicized terms, write the names of the following organs in the table and identify which principle body cavity each would be found. ***Abdominopelvic, Cranial, Spinal, Thoracic***

Brain	Spinal cord	Lungs	stomach
Liver	Urinary Bladder	Heart	Testes
Ovary	Esophagus	Uterus	Pancreas
Rectum	trachea	Small intestines	

2. Copy each of the following statements on your own paper and fill in the appropriate word choice.

(superior or inferior)

1. Abdomen_____vs. pectoral region_____
2. Oral region_____vs. nose_____
3. Cervical region_____vs. tail bone_____

(anterior/ventral or posterior/dorsal)

1. Nose_____vs. ear_____
2. Knuckles_____vs. palm_____
3. Heel_____vs. toes_____

(medial or lateral)

1. Radius bone_____vs. ulna bone_____
2. Middle toe_____vs. big toe_____
3. Orbital region_____vs. ear_____

(proximal or distal)

1. Fingers_____vs. carpal region_____
2. Upper arm_____vs. clavicle_____
3. Lower leg_____vs. thigh_____

3.

Name the nine Abdominopelvic regions	Name the four Abdominopelvic quadrants
1.	1.
2.	2.
3.	3.
4.	4.
5.	
6.	
7.	
8.	
9.	

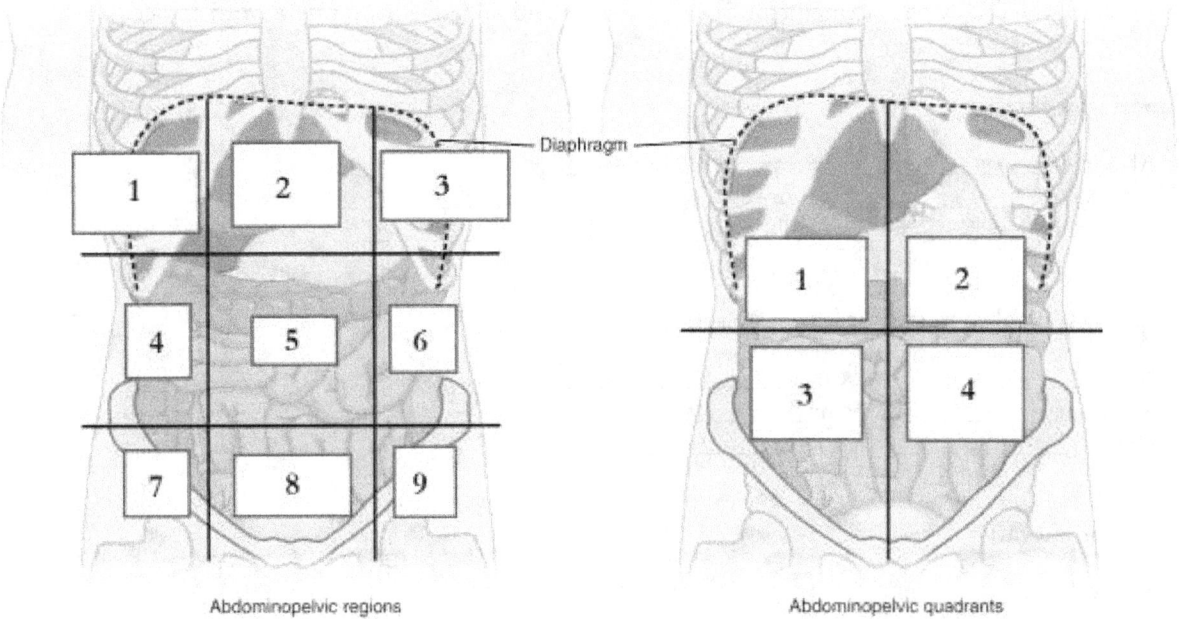

EXERCISE 2.6
MRI

Purpose of exercise: To demonstrate how internal imaging devices are used to identify body structures.

Click on *Exercise 2.6* within your online platform or enter the address below into your web browser:
https://www.nobelprize.org/educational/medicine/mri/game/index.html

EXERCISE 2.7
ANATOMICAL PARTS

Purpose of exercise: To describe the general function of each organ system.

Describe the major function(s) of the eleven organ systems next to its description.

1. Cardiovascular/ Circulatory system:

2. Digestive system:

3. Endocrine system:

4. Integumentary system:

5. Lymphatic system:

6. Muscular system:

7. Skeletal system:

8. Nervous system:

9. Urinary system:

10. Reproductive:

11. Respiratory:

CHEMICAL BASIS OF LIFE
LAB 3

CRASHCOURSE VIDEO(S)

Click on the video embedded within your online platform or enter the address below into your web browser:
1. https://youtu.be/QXT4OVM4vXI
2. https://youtu.be/PVL24HAesnc
3. https://youtu.be/HVT3Y3_gHGg
4. https://youtu.be/LS67vS10O5Y
5. https://youtu.be/H8WJ2KENlK0
6. https://youtu.be/fR3NxCR9z2U
7. https://youtu.be/kb146Y1igTQ

(Please make sure to watch the videos before continuing)

DEFINING KEY TERMS:

1. Acid:

2. Anion:

3. Atom:

4. Atomic Number:

5. Atomic Weight:

6. Base/ Alkaline:

7. Bond:

8. Buffering Systems:

9. Cation:

10. Compound:

11. Covalent Bonds:

12. Decomposition reactions:

13. Electrolytes:

14. Electron shell (energy level):

15. Electron:

16. Element:

17. Homeostasis:

18. Hydrogen Bonds:

19. Inorganic Substances:

20. Ionic Bonds:

21. Ions:

22. Isotopes:

23. Matter:

24. Metabolism:

25. Molecule:

26. Neutron:

27. Octet Rule:

28. Organic Substances:

29. pH:

30. Polarity:

31. Proton:

32. Solute:

33. Solution:

34. Solvent:

35. Synthesis reactions:

36. Valence shell:

37. Vaporization:

Living organisms are composed of both inorganic and organic matter. Inorganic Substances include water, oxygen, carbon dioxide, and inorganic salts. Organic Substances, commonly referred to as macromolecules, in cells include carbohydrates, lipids, proteins, and nucleic acids. It is very important for these Healthcare professionals to learn and understand how these substances and other chemicals react with the human body so that they can take appropriate action when giving care to patients. Healthcare professionals commonly deal with therapeutic drugs as well, which are basically chemical compounds. The manner in which a drug reacts with the human body and with other drugs depends on its chemical composition and therefore on its properties. If a healthcare professional has a limited amount of information about what a person ate or drank, he or she with knowledge of chemical reactions can quickly guess what may have happened to explain the symptoms.

One situation where knowledge of chemical reactions is useful to healthcare professionals is when combining different drugs. Another situation where such knowledge is useful is in safety, handling, and storage procedures for drugs. This can be beneficial or harmful. Another situation is in diagnosing symptoms. The drugs can interact with each other or one drug can produce a by-product which adversely affects the other drug. The chemical reactions that lie behind these interactions are very important to healthcare professionals.

EXERCISE 3.1
IONIC AND COVALENT BONDS

Purpose of exercise: To distinguish the different properties of covalently bonded compounds and ionically bonded compounds.

This virtual lab will describe the different properties of covalently bonded compounds and ionically bonded compounds. Then, you will join in on an actual lab video to see how compounds behave differently depending on whether they are made up of covalent bonds or ionic bonds. The table compares some general properties of each of these bonds.

Ionic Compounds	**Covalent Compounds**
High Melting and Boiling Point	Low Melting and Boiling Point
Harder and Inflexible	Softer and mushier
Less flammable	More flammable
Soluble in water	Insoluble in water
Conducts electricity in water	Doesn't conduct electricity in water

Now that you know the general properties of these bonds, you can watch an online experiment where they measure these properties in compounds to help you identify which compounds contain either covalent or ionic bonds.

This video tests the melting point, solubility, and conductivity of various compounds to determine whether the compounds contain either covalent or ionic bonds.

Click on *Exercise 3.1* within your online platform or enter the address below into your web browser:
https://youtu.be/qZW_6Ev0AXc
Use the table below to indicate the different properties observed in the experiment.

Type of Compound	Melting Point	Solubility in Water	Conductivity
Sucrose ($C_{12}H_{22}O_{11}$)			
Sodium Chloride (NaCl)			
Oxalic Acid ($C_2H_2O_4$)			
Cobalt (II) Sulfate ($CoSO_4$)			
Nickel (II) Chloride ($NiCl_2$)			
Starch ($C_6H_{10}O_5$)			

EXERCISE 3.2
PROPERTIES OF INORGANIC SUBSTANCES: HIGH HEAT OF VAPORIZATION

Purpose of exercise: To demonstrate the high heat of vaporization property associated with water.

This is a hands-on experiment, so you will need to acquire the items listed in the material section of this exercise to perform. If you cannot get the materials, please make predictions about what you think will happen in the experiment. This will also vastly increase the amount that you'll learn.

Materials:
- Paper Towels
- Water
- Timer or Stopwatch
- 70% Isopropyl Alcohol

Procedure:
1. Take a paper towel and rub water on the inside of your forearm.
2. Let your moistened arm air dry (room temperature) inside the lab, and observe how long it takes for the water to completely evaporate so that your arm is completely dry.
3. Repeat this activity with alcohol. Record your data in the data chart at right.

Water Time to dry	Alcohol Time to dry

Question:
Which liquid appears to dry faster, and why?

EXERCISE 3.3
MEASUREMENT OF PH

Purpose of exercise: To determine the pH of several different substances through the use of acid-base indicators.

Click on *Exercise 3.3 (a)* within your online platform or enter the address below into your web browser:
https://www.brainpop.com/games/virtuallabsphscaleandmetercalibration/

Please read the instructions provided on the website before starting the exercise.

Read the introduction and answer the following questions in your laboratory manual:

1. What does the pH of a solution measure?

2. How does pH paper help to determine the pH?

3. What is the range of the pH scale?

4. What is the pH of an acid?

5. What is the pH of a base?

6. What is the pH of a neutral substance?

Write down the pH and the color of the paper for the following substances you tested. Decide if the substance is acidic/basic or neutral.

	Substance	pH	Color	Acid/Base/Neutral
1.	Lemon			
2.	Coffee			
3.	Water			
4.	Milk of Magnesia			

Click on *Exercise 3.3 (b)* within your online platform or enter the address below into your web browser:
http://phet.colorado.edu/sims/html/ph-scale/latest/ph-scale_en.html

This website link routes you to a virtual pH scale. Click on the tab labeled, **"Micro"**

Experiment with each solution. Notice what happens to the bar graph as the pH changes. Record the following:

	pH	$[H_3O^+]$	$[OH^-]$
1. drain cleaner			
2. hand soap			
3. blood			
4. spit			
5. water			
6. milk			
7. coffee			
8. beer			
9. soda pop			
10. vomit			
11. battery acid			

EXERCISE 3.4
PROPERTIES OF ORGANIC SUBSTANCES

Purpose of exercise: To test for the presence of carbohydrates, proteins, and lipids in food by using chemical reagents.

There are four broad classes of macromolecules (carbohydrates, proteins, lipids, and nucleic acids) that can be found in living systems. Each type of macromolecule has a characteristic structure and function in humans. Healthcare professionals have designed tests to determine the presence of these nutrients in food. We can use chemical reagents that react in predictable ways in the presence of these nutrients. Such information may help to maintain a balance intake of the macromolecules.

Use the information from your lecture book and the reading below to help you answer the questions in this exercise.

Tests: Chemicals are used many times as indicators - that is, they show us if the presence of specific macromolecules can be found in given substances. This lab focuses on identifying three of the four macromolecules we have studied (carbs, lipids, proteins) using INDICATORS**. Read the procedures for the Macromolecule Lab and complete the indicator table below.

**INDICATORS are chemicals that react or change color in the presence of another compound. We use them to test for the presence (or absence!) of particular compounds.

Table 1. A list of the indicators (detection reagents) used to reveal the presence of specific macromolecules

Test (Procedure)	Structure/Molecule Detected:	Structure/Molecule Found in:
Benedict's	Reducing Sugars (sugars with a free aldehyde or ketone group; typically mono disaccharides)	
Iodine	Starch	Carbohydrates
Stain	Water insoluble substances	Lipids
Biuret's	Peptide Bonds	Protein

Table 2. Chemical explanations for he colorimetric changes observed in macromolecule detection tests.

Detection Reagent	Explanation of Detection
Benedict's	Contains Copper Sulfate. Copper binds to oxygen in the free aldehyde or ketone group and the Copper Oxide that is formed transmits a brown color
Iodine	Iodine interacts with and binds to a structure in the starch molecule, the new structure transmits a dark bluish black color
Stain	Water insoluble substances interact with other water insoluble substances. Does it dissolve in water?
Biuret's	Contains Copper Sulfate and Sodium Hydroxide. Copper Sulfate actively binds to the peptide bonds found in proteins, and the structure formed transmits a violet color in an alkaline (basic) environment, which is provided by the presence of the NaOH.

Questions
Answer the following questions

1. What are macromolecules and explain why they are important?

2. Names four detection reagents used to test the presence of macromolecules in foods.

3. Monosaccharides and polysaccharides are two classes of _____.

4. Long chains of amino acids make up _____ and contain the element _____ which is unique to this macromolecule.

5. Fats like triglycerides are the macromolecule _____.

6. Glucose is a simple sugar called a _____, whereas starch contains chains of glucose and is a _____.

7. For each of the following tests, please circle which substance would give a positive result:
    a. Benedict's test -   glucose   tap water   oil   starch   protein
    b. Iodine test -   glucose   tap water   oil   starch   protein
    c. Sudan III test -   glucose   tap water   oil   starch   protein
    d. Biuret test -   glucose   tap water   oil   starch   protein

8. What elements are present in all macromolecules?

9. Are all macromolecules found in food? Explain.

10. What element(s) are present in proteins, but are **NOT** found in carbohydrates or lipids?

11. Biurets solution reacts with the peptide bonds in the polypeptide chains. A purple colored complex is formed. What macromolecule has peptide bonds?

12. Explain why the structure of a macromolecule is so important to function. Give an example.

13. Name the macromolecule that each food would contain.

    a. Water

    b. Oil

    c. Milk

    d. Oatmeal

    e. Apple Juice

14. How do you know from the test if a macromolecule is present?

15. ID the test:
    a. A cloudy, orange color shows a positive result for the _____ test.
    b. If a solution contains macromolecules that test positive for the _____ test, it will produce a violet color.
    c. The reagent used in the _____ will turn the dark red solution reddish-orange

EXERCISE 3.5
ENZYMES

Purpose of exercise: To demonstrate how enzymes control metabolic reactions and how metabolic pathways are regulated.

Click on *Exercise 3.5 (a)* within your online platform or enter the address below into your web browser:
https://youtu.be/UVeoXYJlBtI

Click on *Exercise 3.5 (b)* within your online platform or enter the address below into your web browser:
https://youtu.be/pVoytz_3H_s

Click on *Exercise 3.5 (c)* within your online platform or enter the address below into your web browser:
https://youtu.be/ueup2PTkFW8

Watch all of the animations and answer the following questions.

1. What does it mean that enzymes are specific?

2. Enzymes are reusable. So, would you expect to find large amounts in your cells or small amounts?

3. Define denaturation. List two factors that affect enzyme function.

4. _____ catalyze reactions by lowering the _____ necessary for a reaction to occur.

5. _____ catalyze reactions by lowering the _____ necessary for a reaction to occur. The molecule that an enzyme acts on is called the _____. In an enzyme-mediated reaction, _____ molecules are changed, and _____ is formed. The _____ molecule is _____ after the reaction, and it can continue _____.

6. Draw a reaction using shapes, labeling the substrate, enzyme, active site on the enzyme, enzyme-substrate complex, and the products.

EXERCISE 3.6
REVIEW QUESTIONS

Please continue by answering the questions below.

1. For the elements chlorine and calcium, the chemical symbols are _____ and For the elements iron and iodine, the chemical symbols are _____ and _____.
2. For the elements potassium and phosphorus, the chemical symbols are _____ and _____.
3. An ionic bond is formed when atoms gain or lose _____.
4. When atoms gain or lose electrons, a(n) _____ bond is formed.
5. An atom that has lost or gained electrons is called a(n) _____.
6. The number of positive or negative charges an ion has is called its _____.
7. An anion is an ion with a(n) _____ charge.
8. A cation is an ion with a(n) _____ charge.
9. The bond between sodium and chloride in a molecule of NaCl is a(n) _____ bond.
10. A bond in which electrons are shared between two atoms is a(n) _____ bond.
11. The weak bonds that help maintain the 3-D shape of proteins and nucleic acids are _____ bonds.
12. The subunits of DNA and RNA are called _____.
13. A nucleotide consists of a phosphate group, a(n) _____, and a(n) _____.
14. The subunits of a molecule of glycogen are molecules of _____.
15. Two polysaccharides made of glucose are _____ and _____.
16. Amino acids are the subunits of _____.
17. The bonds between the amino acids in a protein are _____ bonds.
18. The body as a whole changes temperature slowly because the body is mostly _____.
19. While sweating, the body loses heat by the process of _____.
20. The evaporation of sweat is a mechanism for the loss of _____.
21. A disadvantage of sweating is that it may lead to _____.
22. The excretion of waste products in urine depends on the _____ function of water.
23. The synovial fluid of joints is an example of the _____ function of water.
24. The genetic material of the chromosomes of a cell is _____.
25. DNA is the genetic material in the _____ of a cell.

26. Glycogen is the storage form for excess _____.
27. Phospholipids are present in all human cells as part of _____.
28. The precursor molecule for the steroid hormones is _____.
29. Cells in the ovaries use cholesterol to synthesize _____.
30. Cells in the testes use cholesterol to synthesize _____.
31. The energy products of cell respiration are _____ and _____.
32. The purpose of cell respiration is to produce _____ from _____.
33. Biologically useful energy is released in cell respiration in the form of _____.
34. The waste product of cell respiration is _____.
35. The accumulation of carbon dioxide will cause the _____ of body fluids to decrease.
36. The trace element _____ is part of some thyroid hormones.
37. On the pH scale, acids are indicated by numbers _____.
38. On the pH scale, bases are indicated by numbers _____.
39. The active site of an enzyme is the place where the _____ molecule(s) fit.
40. A high fever may change the shape of enzymes; that is, enzymes become _____.

CELL STRUCTURE AND FUNCTIONS
LAB 4

CRASHCOURSE VIDEO(S)

Click on the video embedded within your online platform or enter the address below into your web browser:
1. **https://youtu.be/cj8dDTHGJBY**
2. **https://youtu.be/dPKvHrD1eS4**
3. **https://youtu.be/L0k-enzoeOM**
4. **https://youtu.be/00jbG_cfGuQ**
5. **https://youtu.be/8kK2zwjRV0M**
6. **https://youtu.be/CBezq1fFUEA**

(Please watch the videos below before continuing)

DEFINING KEY TERMS:

1. Active transport:

2. Allele:

3. Anabolism:

4. Anticodon:

5. ATP:

6. Catabolism:

7. Cell Cycle:

8. Cell Theory:

9. Codon:

10. Complementary base pairing:

11. Diffusion:

12. Diploid

13. DNA Extraction:

14. DNA:

15. Dominant Trait:

16. Enzyme:

17. Extraction:

18. Filtration:

19. Genetics:

20. Genotype:

21. Gradient:

22. Haploid:

23. Heredity:

24. Heterozygous:

25. Hydrophilic:

26. Hydrophobic:

27. Meiosis:

28. Metabolism:

29. Mitosis:

30. Osmosis:

31. Passive transport:

32. Phenotype:

33. Protein Synthesis:

34. Recessive Trait:

35. RNA:

36. Selectively permeable:

37. Transcription:

38. Translation:

EXERCISE 4.1
CELL ORGANELLES

Purpose of exercise: To describe each kind of cytoplasmic organelle and explain its function.

Write the function of the cell organelles in the space provided below:

CELL COMPONENT	FUNCTION(S)
Plasma membrane	
Cytoplasm	
Ribosomes	
Rough Endoplasmic Reticulum	
Smooth Endoplasmic Reticulum	
Golgi	
Vesicles	
Mitochondria	
Lysosomes	
Peroxisomes	
Centrosomes	
Cilia	
Flagella	
Microvilli	
Cytoskeleton	
Nucleus	
Nucleolus	
Chromosome	

EXERCISE 4.2
PARTS OF THE CELL

Purpose of exercise: To observe the general characteristics of a composite cell.

Click on *Exercise 4.2 (a)* within your online platform or enter the address below into your web browser:
http://learn.genetics.utah.edu/content/cells/scale/

This will allow you to compare the cell size and scale with other structures. Please read and follow the instructions provided on the website. After you have used the previous hyperlink move to the next.

Click on *Exercise 4.2 (b)* within your online platform or enter the address below into your web browser:
http://www.cellsalive.com/cells/cell_model_js.htm

Choose the Animal Cell from the Eukaryotic Organelle Cell Model choices given. You are viewing an interactive cell model. Use the model to help reinforce your understand of the cell.

Label the parts of the cell in your laboratory manual.

EXERCISE 4.3
CELL Explorer: The Animal Cell

Purpose of exercise: To interact with the cytoplasmic organelles of the composite cell.

Click on *Exercise 4.3 (a)* within your online platform or enter the address below into your web browser:
http://learn.genetics.utah.edu/content/cells/insideacell/

Please read and follow the instructions provided on the website. After you have used the previous hyperlink move to the next.

Click on *Exercise 4.3 (b)* within your online platform or enter the address below into your web browser:
https://biomanbio.com/HTML5GamesandLabs/Cellgames/cellexplorerpagehtml5.html

Please read and follow the instructions provided on the website. After you have used the previous hyperlink move to the next.

Click on *Exercise 4.3 (c)* within your online platform or enter the address below into your web browser
https://biomanbio.com/GamesandLabs_old/Cellgames/cellcraft.html

Please read and follow the instructions provided on the website.

EXERCISE 4.4
CHARACTERISTICS OF THE PLASMA MEMBRANE

Purpose of exercise: To demonstrate the structure and functional properties of the plasma membrane of a cell.

This is a hands-on experiment, so you will need to acquire the items listed in the material section of this exercise to perform. I would also suggest that you perform this exercise outside.

Plasma membranes are similar to bubbles. They are fluid, flexible, and can self-repair. The membrane plasma membrane is a double layer of composed of phospholipids. Phospholipids have a "head" that is attracted to water, and a "tail" that is repelled by water. If you place phospholipids in water, they quickly form a double layer with the heads facing out on both sides.

A soap molecule is very similar to a phospholipid. The "head" of a soap molecule is charged (ionic) and attracts to water molecules, which have regions of positive and negative charge (polar). The hydrocarbon tail of the soap molecule is not charged and is repelled by water's polarity. The hydrocarbon tail of soap mixes with and dissolves in other hydrocarbons, like oils and fats, while the head region grabs a hold of passing water molecules and follows them down the drain. The surface of a bubble has three layers. The middle layer is a thin film of water. On both sides of this film is a layer of soap molecules with hydrophilic heads oriented toward the water film and hydrophobic tails pointing away.

Materials
- 44 ounce cup
- 32 ounces of water
- 7 tablespoons of liquid dish soap

- 2 tablespoons of corn syrup
- 5 bendable drinking straws
- Standard-sized 18" x 26" x 1" Aluminum Bun Pan/Sheet Pan

Procedure
1. Create the bubble solution by mixing the water, soap, and corn syrup in the 44 ounces cup
2. Create a bubble frame by ending 4 straws at elbows.
3. Flatten the shorter ends of straws and fold flatted surface in the middle

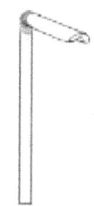

4. Connect straws together by inserting short ends into long ends to create a square.

5. Place bubble frame into shallow tray
6. Add bubble solution to slightly cover bubble frame.

Membranes are Fluid and Flexible
Plasma membranes are not stationary, they bend and flex in order to adapt to changing conditions.
1. Lift bubble frame out of solution so that a thin film spans across frame.
2. Tilt the frame back and forth and observe the surface of the film.
3. Notice the swirl of color as the light reflects off the film. Molecules in the plasma membrane move about in a similar fashion.
4. Hold the frame by the edges and rotate the sides in opposite directions. Notice the elasticity of the film

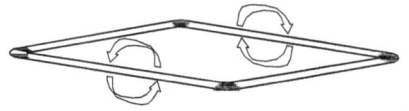

5. Hold the bubble film parallel to the floor and gently move the frame up and down until the surface begins to bounce up and down.
6. Like the bubble film, plasma membranes can flex without breaking.

Membranes Can Self-Repair
Attraction between phospholipids allows plasma membranes to repair small breaks in the bilayer.

1. Lift bubble frame out of solution so that a thin film spans across frame.
2. Cover the surface of your finger or extra straw in bubble solution.

3. Slowly push finger or straw through film. Film should allow finger to pass without breaking.
4. Remove finger from film. Film should repair itself.
5. Try the same procedure with your entire hand.
6. Like the bubble layer, cell membranes can spontaneously repair small tears in the lipid bilayer.

Eukaryotic Cells Feature Membrane Bound Organelles

The membranes surrounding organelles in Eukaryotic cells feature a phospholipid bilayer like the one found in the outer "plasma" membrane.

1. Place the tip of a clean straw into the bubble solution in the Sheet Pan.
2. Gently blow on the other end of the straw to create a bubble.
3. Slowly lift the tip of the straw out of the liquid while continuing to fill the bubble with air.
4. Allow the bubble to grow to a size of about 6" wide.
5. Return the tip of the straw back into the bubble solution and try to create a smaller bubble inside the larger bubble.
6. Notice how the smaller bubble creates a compartment of air that is contained within but separated from the air of the larger bubble.
7. In a similar fashion, Eukaryotic cells feature membrane bound organelles that create specialized compartments within a single cell. The primary structure of the outer cell membrane as well as the membranes that enclose organelles is a double layer of phospholipids known as a phospholipid bilayer.

Gap Junctions Aid Transport between Animal Cells

Gap junctions form between neighboring animal cells, allowing their cytoplasm to connect directly. This provides a quick way to move material between two cells.

1. Place the tip of a clean straw into the bubble solution in the Sheet Pan.
2. Gently blow on the other end of the straw to create a bubble.
3. Lift the straw out of the liquid and continue to blow air into the bubble.
4. Continue to gently blow air into the bubble as you slowly pull out the straw.
5. As the tip of the straw leaves the bubble, you should be able to briefly observe a small tunnel existing between the bubble and straw tip.
6. This tunnel is allowing air to freely move between the straw and the bubble without breaking the bubble film.
7. In the same manner, gap junctions allow for the rapid transit of ions and small molecules between adjacent animal cell membranes.

EXERCISE 4.5
CELL DEFENSE: THE PLASMA MEMBRANE

Purpose of exercise: To demonstrate the structure and functional properties of the plasma membrane of a cell.

Click on *Exercise 4.5* within your online platform or enter the address below into your web browser:
https://biomanbio.com/GamesandLabs_old/Cellgames/celldefense.html

Please read and follow the instructions provided on the website

EXERCISE 4.6
MOVEMENTS INTO AND OUT OF THE CELL

Purpose of exercise: To demonstrate the processes of diffusion, osmosis, passive transport, and active transport in animal cells.

Click on *Exercise 4.6* within your online platform or enter the address below into your web browser:
http://www.wiley.com/legacy/college/boyer/0470003790/animations/membrane_transport/membrane_transport.htm

Click on each of the green tabs located under Cellular Transport in the Interactive animations. Sketch the illustration given when you click on the following green tabs:

1. Membranes

2. Diffusion/osmosis

3. Active transport

EXERCISE 4.7
CELLS GAIN OR LOSS OF WATER

Purpose of exercise: To demonstrate how osmosis and tonicity affect the cell.

Click on *Exercise 4.7* within your online platform or enter the address below into your web browser:
http://www.phschool.com/science/biology_place/labbench/index.html

Below the **LabBench Activities** heading, click on **Lab 1: Diffusion & Osmosis**. Read and follow the instructions of the lab activity. Click **Closer Look** or **Next Concept** to continue through the activity until you reach and complete the page with the sub-heading **Water Potential**. This will be listed below the **LabBench Activity** heading. Please stop here.

EXERCISE 4.8
CELL DIVISON: CELL CYCLE

Purpose of exercise: To demonstrate and explain how cells divide.

Click on *Exercise 4.8 (a)* within your online platform or enter the address below into your web browser:
https://www.nobelprize.org/educational/medicine/2001/

Once you have completed the "Control of the Cell Cycle game" move to the next activity.

Click on *Exercise 4.8 (b)* within your online platform or enter the address below into your web browser:
https://youtu.be/woD6zvp-4E8

(Please watch the video before labeling the parts of the cell cycle)

1. Label the parts of the cell cycle

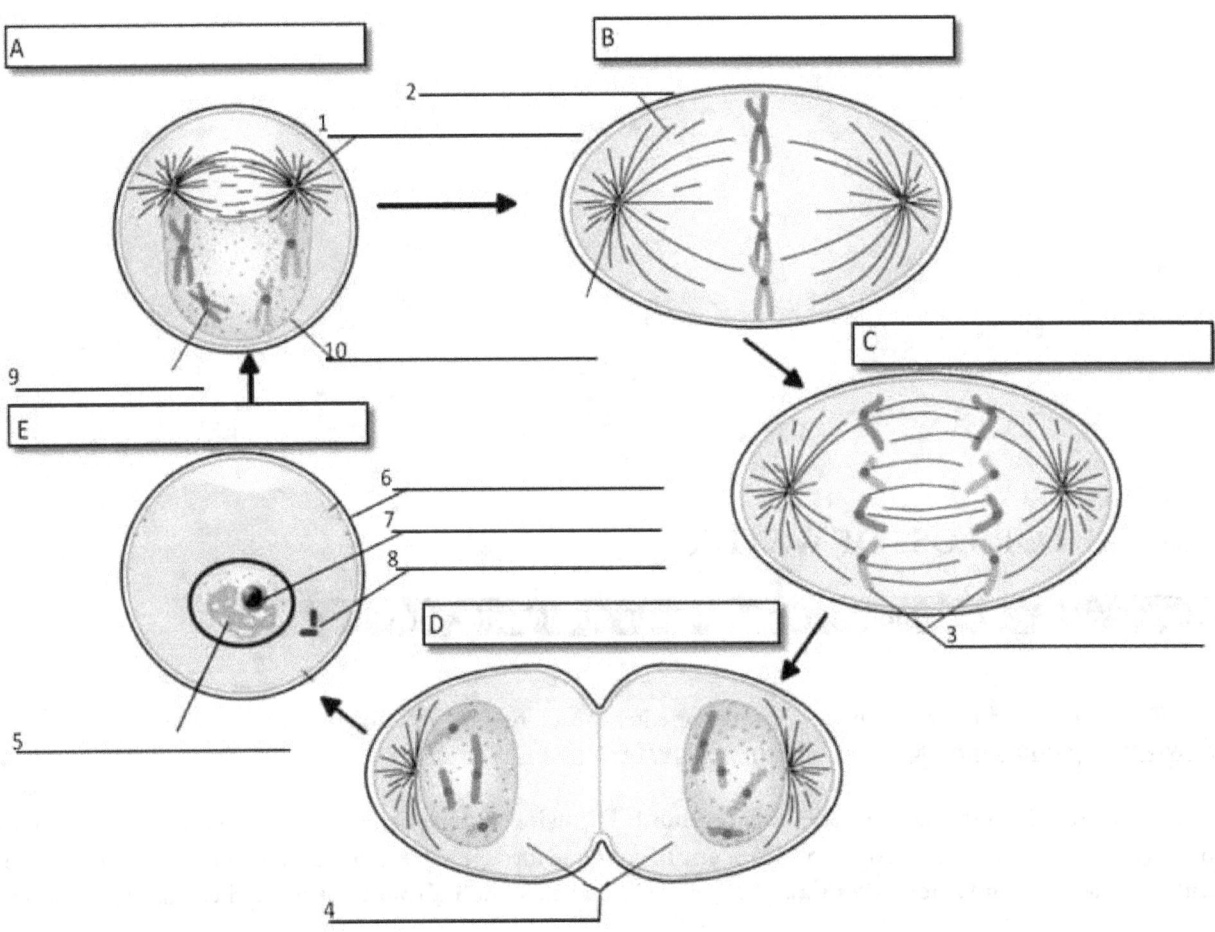

Complete the table by describing the events of the cell cycle.

Name of Phase	Description of Events
Interphase	
Prophase	
Metaphase	
Anaphase	
Telophase	
Cytokinesis	

EXERCISE 4.9
CELL DIVISON: MITOSIS

Purpose of exercise: To correctly identify and draw the four stages of mitosis using microscope slide images of the whitefish blastula.

Human chromosomes are not clearly visible at higher power magnification. So, for student purposes, whitefish blastula are used. The blastula is an early stage of embryo development and represents a period in the organism's life when most of the cells are constantly dividing. The dividing cell have very large and easily seen chromosomes, so it's easy to find cells in each stage of mitosis.

For a great review of mitosis, click on *Exercise 4.9 (a)* within your online platform or enter the address below into your web browser:
https://youtu.be/C6hn3sA0ip0

Mitosis of Whitefish Blastula
(Please use one of the two addresses provided below.)

Click on *Exercise 4.9 (b)* within your online platform or enter the address below into your web browser:
http://www.sciencegeek.net/Biology/Mitosis/Mitosis2.shtml

-or-

Click on *Exercise 4.9 (c)* within your online platform or enter the address below into your web browser:
http://science.jburroughs.org/resources/mitosis/interphasewfb.html

1. Scroll through the slides to see the best examples of the four stages of mitosis. When you are confident that you have identified each stage, perform a sketch of each stage in the table below.

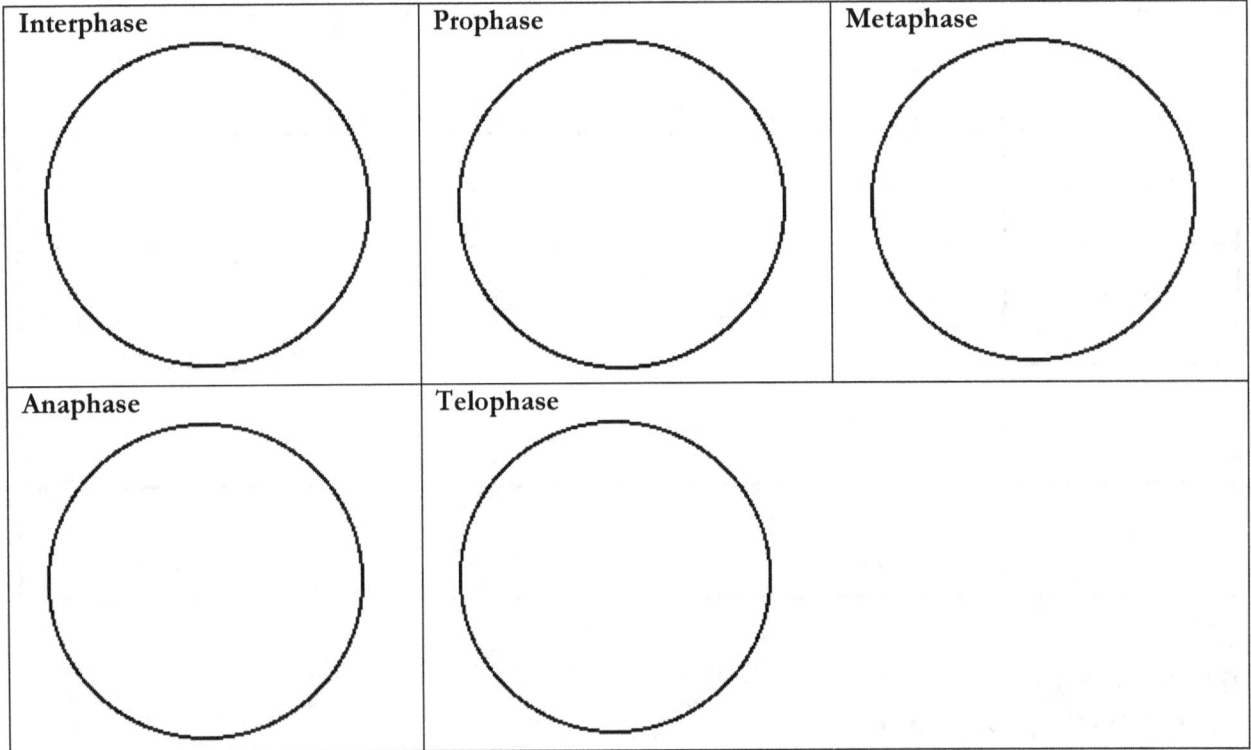

EXERCISE 4.10
STEM CELLS

Purpose of exercise: To send activating signals to stem cells and watch them get to work.

Click on *Exercise 4.10 (a)* within your online platform or enter the address below into your web browser:
http://learn.genetics.utah.edu/content/stemcells/scintro/

Click on *Exercise 4.10 (b)* within your online platform or enter the address below into your web browser:
http://learn.genetics.utah.edu/content/stemcells/sctypes/

Please read and follow the instructions provided on the website

EXERCISE 4.11
CLONING

Purpose of exercise: To demonstrate the process of cloning using a virtual mouse specimen.

Click on *Exercise 4.11 (a)* within your online platform or enter the address below into your web browser:
http://learn.genetics.utah.edu/content/cloning/whatiscloning/

Please read and follow the instructions provided on the website. After you have used the previous hyperlink move to the next.

Click on *Exercise 4.11 (b)* within your online platform or enter the address below into your web browser:
http://learn.genetics.utah.edu/content/cloning/clickandclone/

Please read and follow the instructions provided on the website

EXERCISE 4.12
CELLULAR RESPIRATION

Purpose of exercise: To explain how the reactions of cellular respiration release chemical energy.

Answer the following questions:

1. What are the 3 phases of the cellular respiration process?

2. Where in the cell does the glycolysis part of cellular respiration occur?

3. Where in the cell does the Krebs (Citric Acid) cycle part of cellular respiration occur?

4. Where in the cell does the electron transport part of cellular respiration occur?

5. How many ATP (net) are made in the glycolysis part of cellular respiration?

6. How many ATP are made in the Kreb's cycle part of cellular respiration?

7. How many ATP are made in the electron transport part of cellular respiration?

8. In which phase of cellular respiration is carbon dioxide made?

9. In which phase of cellular respiration is water made?

10. In which phase of cellular respiration is oxygen a substrate?

11. In which phase of cellular respiration is glucose a substrate?

12. On average, how many ATP can be made from each NADH during the electron transport process?

13. On average, how many ATP can be made from each FADH₂ during the electron transport process?

14. What would happen to the cellular respiration process if the enzyme for one step of the process were missing or defective?

15. What is the overall reaction for lactic acid fermentation?

16. Only a small part of the energy released from the glucose molecule during glycolysis is stored in ATP. In what form is the rest of the energy released? (HINT: It is a product in the overall reaction for cellular respiration.)

17. Write the complete overall chemical equation for cellular respiration using chemical symbols instead of words:

Match the Letter in the Diagram with the Label:
(You can use them MORE THAN ONCE or NOT AT ALL)

1. _____ Place where glycolysis happens

2. _____ Place where enzymes for the Electron Transport Chain are located

3. _____ Place that fills with H⁺ ions as electrons move down the Electron Transport Chain

4. _____ Place where ADP and P join to make ATP

5. _____ Place where oxygen acts as the final electron acceptor to make water

Cellular Respiration Vocabulary Review

1. _____ is a 6 carbon molecule that is produced first when acetyl-CoA joins with a 4 carbon molecule to enter the Krebs cycle.
2. _____ is the process of splitting a glucose molecule into 2 pyruvic acid molecules.
3. The molecule used by cells to store and transfer energy is _____
4. Glycolysis happens outside the mitochondria in the _____ of the cell.
5. _____ happens when oxygen is present and includes glycolysis, Krebs cycle, and Electron Transport.
6. This describes a process that requires oxygen = _____
7. This atmospheric gas is required for aerobic respiration = _____
8. This describes a process that does NOT require oxygen; it means "without air"
    a. = _____
9. The _____ cycle breaks down pyruvic acid into carbon dioxide and produces NADH, FADH₂, and ATP.
10. The NADH and FADH2 produced during the Krebs cycle pass their electrons down the _____ chain to produce ATP.
11. The passage of H⁺ ions through _____ causes it to spin and produce ATP.
12. This 3 carbon molecule is produced during glycolysis when glucose splits in half
    a. = _____
13. Cell organelle which acts as the cell's power plant to burn glucose and store energy as ATP
    a. = _____
14. If oxygen is NOT present, glycolysis is followed by _____.
15. This molecule has the formula $C_6H_{12}O_6$ and is split in half during glycolysis
    a. = _____
16. The carbon atoms in pyruvic acid end up as _____ in the atmosphere following the Krebs cycle.

17. The folded inner membranes inside a mitochondrion are called _____
18. This molecule reacts with pyruvic acid to release CO_2, produce NADH, and acetyl-CoA.
    a. = _____
19. _____ forms when Coenzyme A attaches to two carbons from pyruvic acid.
20. _____ is the storage form of glucose used by human cells which can be broken down for energy when glucose is used up.

EXERCISE 4.13
DOUBLE HELIX: (DNA)

Purpose of exercise: To observe the structure of DNA.

Click on *Exercise 4.13 (a)* within your online platform or enter the address below into your web browser:
http://www.hhmi.org/biointeractive/double-helix

Please watch the video provided on the website. After you have used the previous hyperlink move to the next.

Click on *Exercise 4.13 (b)* within your online platform or enter the address below into your web browser:
http://www.hhmi.org/biointeractive/chemical-structure-dna

EXERCISE 4.14
DOUBLE HELIX: (DNA)

Purpose of exercise: To build a DNA molecule.

Click on *Exercise 4.14* within your online platform or enter the address below into your web browser:
http://learn.genetics.utah.edu/content/basics/builddna/

Please read and follow the instructions provided on the website.

EXERCISE 4.15
GEL ELECTROPHORESIS

Purpose of exercise: To sort and measure DNA strands by running your own gel electrophoresis experiment.

Click on *Exercise 4.15* within your online platform or enter the address below into your web browser:
http://learn.genetics.utah.edu/content/labs/gel/

Please read and follow the instructions provided on the website.

EXERCISE 4.16
PCR (SHORT FOR POLYMERASE CHAIN REACTION)

Purpose of exercise: To use a simple and inexpensive tool that you can use to focus in on a segment of DNA and copy it.

Click on *Exercise 4.16* within your online platform or enter the address below into your web browser:
http://learn.genetics.utah.edu/content/labs/pcr/

Please read and follow the instructions provided on the website.

EXERCISE 4.17
DNA MICROARRAY

Purpose of exercise: To use DNA microarray to investigate the differences between a healthy cell and a cancer cell.

Click on *Exercise 4.17* within your online platform or enter the address below into your web browser:
http://learn.genetics.utah.edu/content/labs/microarray/

Please read and follow the instructions provided on the website

EXERCISE 4.18
GENE THERAPY

Purpose of exercise: To design and test gene therapy treatments.

Click on *Exercise 4.18 (a)* within your online platform or enter the address below into your web browser:
http://learn.genetics.utah.edu/content/genetherapy/intro/

Please read and follow the instructions provided on the website. After you have used the previous hyperlink move to the next.

Click on *Exercise 4.18 (b)* within your online platform or enter the address below into your web browser:
http://learn.genetics.utah.edu/content/genetherapy/doctor/

Please read and follow the instructions provided on the website

EXERCISE 4.19
DEOXYRIBONUCLEIC ACID: (DNA) EXTRACTION

Purpose of exercise: To identify the resources and process of DNA Extraction.

Click on *Exercise 4.19* within your online platform or enter the address below into your web browser:
http://learn.genetics.utah.edu/content/labs/extraction

Please read the instructions provided on the website and answer the questions in your lab manual for this exercise.

1. What are some reasons for extracting DNA from human cells
2. Click on **Start Lab** in the right hand corner.
    a. What are 3 reason why scientists isolate DNA?
3. **Click slide 3.** What structure is found in each of our cells? Which macromolecule is found inside this structure?
4. **Click to slide 4.** What is the X-shaped structure inside the nucleus called? What macromolecule does it contain?
5. **Click to next slide.** Which cells are being removed from the man?
6. List the 4 steps that you will use to purify the DNA from the cheek cells.
7. Click through to the next steps. Look closely at the cotton swab, what do you see covered in it?
8. Click to the next step. What is the purpose of the **lysis** solution?
9. The lysis solution contains detergent. What does the detergent cause the cell to do?
10. Click through to the next steps. What does the salt solution do?
11. Click through to the next steps. What does the centrifuge do?
12. Click through to the next steps. What is the purpose of adding the isopropyl alcohol to the tube?
13. **The white stuff left in the bottom of the tube is the DNA!**

EXERCISE 4.20
RNA

Purpose of exercise: To demonstrate how protein synthesis relies on genetic information.

Click on *Exercise 4.20 (a)* within your online platform or enter the address below into your web browser:
http://www.pbs.org/wgbh/nova/labs/lab/rna/

Please read and follow the instructions provided on the website. If you are asked: **Is GPB your local station?** Click **Yes**. You can sign on as a guest, then click **PLAY GAME** written in the yellow box.

Click on *Exercise 4.20 (b)* within your online platform or enter the address below into your web browser:
http://learn.genetics.utah.edu/content/basics/rna/

Please read and follow the instructions provided on the website. After you have used the previous hyperlink move to the next.

Click on *Exercise 4.20 (c)* within your online platform or enter the address below into your web browser:
http://learn.genetics.utah.edu/content/basics/centraldogma/

Please read and follow the instructions provided on the website.

EXERCISE 4.21
TRANSCRIPTION AND TRANSLATION

Purpose of exercise: To see how cells read the information in a DNA sequence to build a protein.

Click on *Exercise 4.21* within your online platform or enter the address below into your web browser:
http://learn.genetics.utah.edu/content/basics/transcribe/

Please read and follow the instructions provided on the website.

EXERCISE 4.22
TYPES OF PROTEINS

Purpose of exercise: To explore the types of proteins and learn about their varied functions.

Click on *Exercise 4.22* within your online platform or enter the address below into your web browser:
http://learn.genetics.utah.edu/content/basics/proteintypes/

Please read and follow the instructions provided on the website.

EXERCISE 4.23
INHERITANCE OF HUMAN TRAITS

Purpose of exercise: To demonstrate inheritance by illustrating the concepts of genotype and phenotype through human facial features.

Click on *Exercise 4.23 (a)* within your online platform or enter the address below into your web browser:
http://learn.genetics.utah.edu/content/basics/traits/

After you have used the previous hyperlink move to the next.

Click on *Exercise 4.23 (b)* within your online platform or enter the address below into your web browser:
http://learn.genetics.utah.edu/content/basics/inheritance/

Now that you have watched the two videos, please complete this exercise in your laboratory manual.

This lab will require one other person to perform. Please ask someone to assist you.

You and your lab partner represent a couple that each have one dominant and one recessive gene for each facial feature illustrated in this lab; this means you are heterozygous for each trait.

Materials:
- Two pennies
- Colored pencils

Procedure:
1. Obtain a lab partner and the rest of your materials. Decide which of you will contribute the genes of the mother and with will contribute the genes of the father.

2. Find out the sex of your child.
   - Remember your mom's genotype is XX and dad's is XY. So only the father will flip the coin to determine the sex of the child.
   - Heads represents Y sperm, which means the child will be a boy.
   - Tails represents X sperm, which means the child, will be a girl.
3. Give your child a name.
4. Determine the facial features your child will have by flipping the coin as directed by the following pages. For purposes of the rest of this lab activity:
   - Heads will represent the dominant trait shown in capital letters.
   - Tails will represent the recessive trait shown in lowercase letters.
5. On the *Data* table, record the genetic contributions (results from the flips of the coins) in the columns labels Gene(s) from Mother and Gene(s) from Father. Record the actual genetic message in the genotype column, and record the appearance in the phenotype column.
6. When you have determined all the features of your child's face, draw and color the way your child will look.

Facial Features

1. Face Shape   Round (RR, Rr)   Square (rr)

2. Chin Shape   Prominent (PP, Pp)   Weak (pp)

3. Chin Shape II – only if your child's chin is prominent (PP, Pp)
                             Round Chin (RR, Rr)   Square Chin (rr)

4. Cleft Chin   Present (CC, Cc)   Absent (cc)

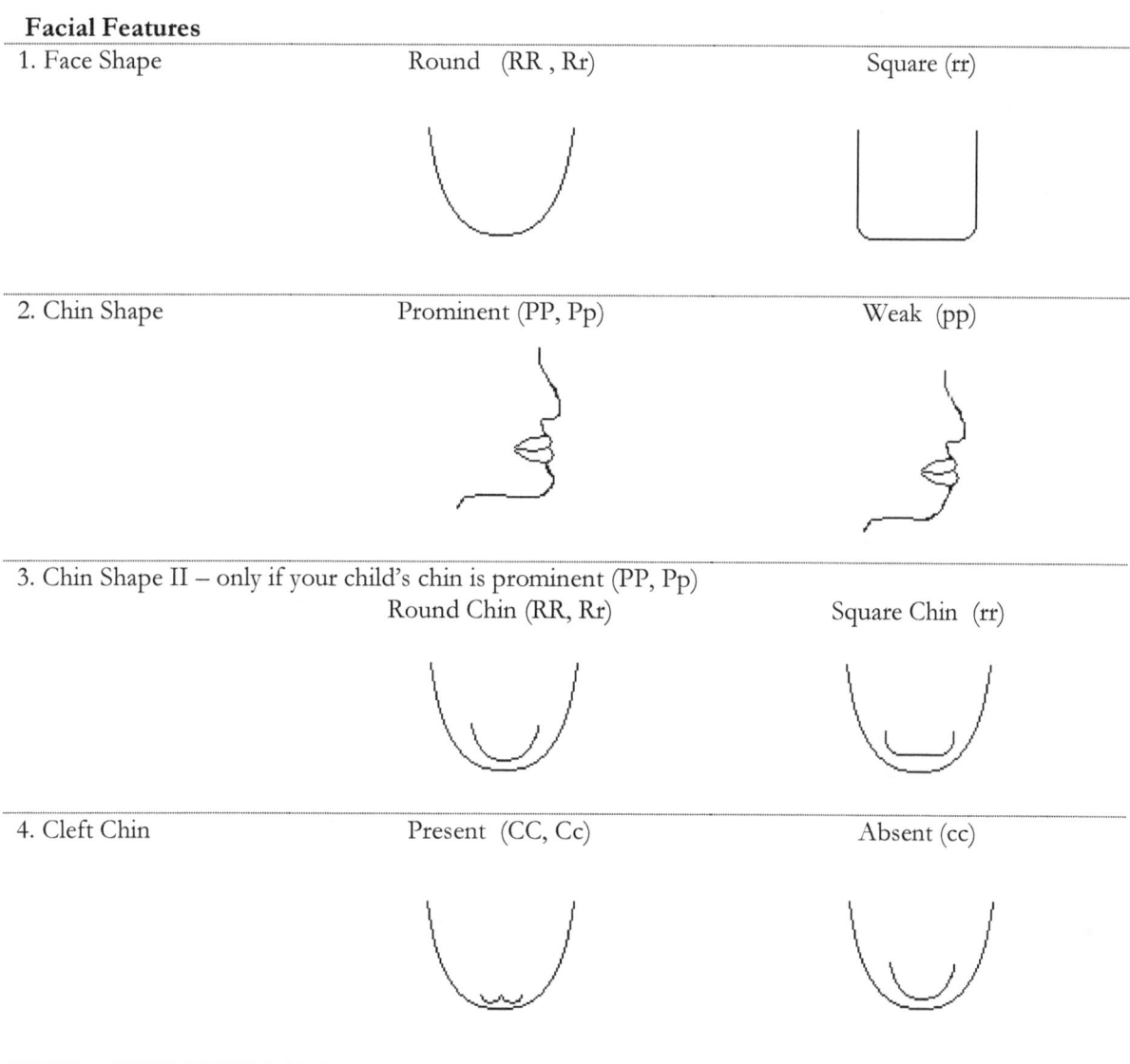

5. Skin Color:

Skin color involves 3 gene pairs. Each parent need to flip the coin 3 times, and record the A, B, and C alleles. For example the result of the first pair of coin flips might be AA, Aa, or aa. Record the first coin flip then do two more alleles B and C. Each capital letter represent an active gene for melanin production (color).

6 capitals	Very dark black skin
5 capitals	Very dark brown
4 capitals	Dark brown
3 capitals	Medium brown
2 capitals	Light brown
1 capitals	Light tan
0 capitals	White

6. Hair Color:

Like skin color hair color is produced by several genes (polygenic or multiple alleles). For the purpose of this activity we will assume that 4 pairs are involved (more are likely). So, each parent will have to flip the coins 4 times for the A, B, C and D alleles. As before, the capital letters (dominant) represent color while the lower case (recessive) represent little or no color.

8 capitals	Black
7 capitals	Very dark brown
6 capitals	Dark brown
5 capitals	Brown
4 capitals	Light brown
3 capitals	Honey blond
2 capitals	Blond
1 capitals	Very light blond
0 capitals	White

7. Red Hair Color

Red hair seems to be caused by a single gene with two alleles:

Dark red (RR)   Light red (Rr)   No red (rr)

Red hair is further complicated by the fact that brown hair will mask or hide red hair color. The lighter the hair color the more the red can show through. If your child has 3 or less capitals (for hair color, see number 6), and RR is tossed your child will have flaming red hair. (Have fun with your colored pencils!)

8. Hair Type: incomplete dominance

Curly (CC)   Wavy (Cc)   Straight (cc)

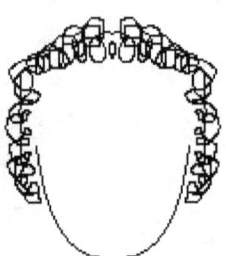  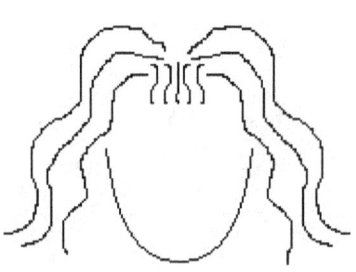

9. Widow's Peak: The hair comes to a point...like Eddie Munster
   Present (WW, Ww)   Absent (ww)

10. Eyebrow Color: incomplete dominance
    Dark (DD)   Medium (Dd)   Light (dd)

11. Eyebrow Thickness: Bushy (BB, Bb)   Fine (bb)

12. Eyebrow Placement: Not connected (NN, Nn)   Connected

13. Eye Color:
    Assume that there are two gene pairs involved, the capital letters represent more color and the lower case, less color. Dark eyes are dominant over light. Assume that there are two layers of color on the iris of the eye. The first alleles (A or a) code for the front of the iris and the second alleles (B or b) code for the back of the iris. Determine the first layer, A, then the second layer, B. In reality eye color is much more complex than this.

AABB	Dark brown
AABb	Dark brown
AaBB	Brown with green flakes
AaBb	Hazel
Aabb	Dark blue
aaBB	Green
aaBb	Grey blue
aabb	Light blue

14. Eye Distance:
    Close together (EE)   Average (Ee)   Far apart (ee)

15. Eye Size: Large (LL)   Average (Ll)   Small (ll)

16. Eye Shape: Almond (AA, Aa)   Round (aa)

17. Eye Tilt: Horizontal (HH, Hh)   Upward slant (hh)

18. Eyelashes: Long (LL, Ll)   Short (ll)

19. Mouth Size: Long (LL)   Average (Ll)   Short (ll)

20. Lip Thickness: Thick (TT, Tt)   Thin (tt)

21. Lip Protrusion: Very protruding (PP)   Slightly protruding (Pp)   Absent (pp)

22. Dimples: Present (PP, Pp)   Absent (pp)

23. Nose Size: Big (BB)   Average (Bb)   Small (bb)

24. Nose Shape: Rounded (RR, Rr)   Pointed (rr)

25. Nostril Shape: Rounded (RR, Rr)   Pointed (rr)

26. Earlobe Attachment: Free (FF, Ff)   Attached (ff)

27. Darwin's Ear Point: Present (PP, Pp)   Absent (pp)

28. Ear Pits: Present (PP, Pp)   Absent (pp)

29. Hairy Ears: This sex-linked and only occurs in males so if your baby girl skip this. If your baby is a boy, only mom flips.
           Present (P)   Absent (p)

30. Freckles on Cheeks:
           Present (PP, Pp)   Absent (pp)

31. Freckles on Forehead:
           Present (PP, Pp)   Absent (pp)

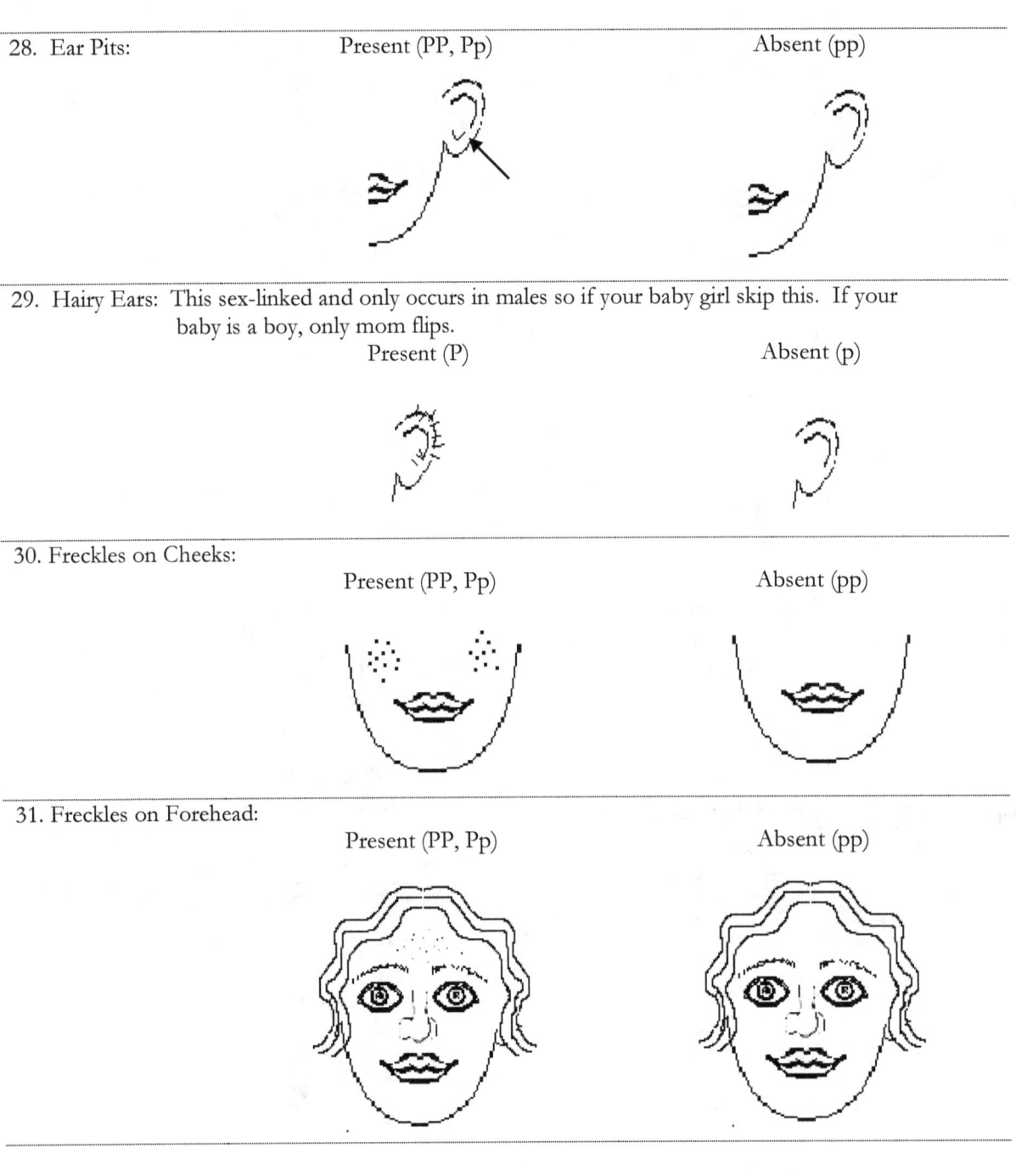

Data

Parent Names: **Baby's Name:** _____

_____

_____

Facial Trait	Genes from Mother	Genes from Father	Genotype	Phenotype
1. Face Shape				
2. Chin Shape				
3. Chin Shape II				
4. Cleft Chin				
5. Skin Color				
6. Hair Color				
7. Red Hair				
8. Hair Type				
9. Widow's Peak				
10. Eyebrow Color				
11. Eyebrow Thickness				
12. Eyebrow Placement				
13. Eye Color				
14. Eye Distance				
15. Eye Size				
16. Eye Shape				
17. Eye Tilt				
18. Eyelashes				
19. Mouth Size				
20. Lip Thickness				
21. Lip Protrusion				
22. Dimples				
23. Nose Size				
24. Nose Shape				
25. Nostril Shape				
26. Earlobe Attachment				
27. Darwin's Ear Point				
28. Ear Pits				
29. Hairy Ears				
30. Freckles on Cheeks				
31. Freckles on Forehead				

Use the color pencils and draw the face of your child in the box.

EXERCISE 4.24
HEREDITY

Purpose of exercise: To explore the genetic basis for human diversity.

Click on *Exercise 4.24* within your online platform or enter the address below into your web browser:
http://labcenter.dnalc.org/labs/mendeliangenetics/mendeliangenetics_d.html

Read and follow the instructions of the lab activity.

EXERCISE 4.25
REVIEW QUESTIONS

1. The stability of cell membranes is increased by the presence of _____.
2. The bilayer of a cell membrane is made of _____.
3. Lipid-soluble materials may diffuse through a cell membrane because of the presence of Channels in a cell membrane are made of _____.
4. Cell membranes allow certain materials in or out; such membranes are said to be _____ permeable.
5. The cell organelles that are the site of protein synthesis are _____.
6. Ribosomes are cell organelles that are the site of _____.
7. The cell organelles that destroy damaged or misfolded proteins are the _____.
8. The Golgi apparatus is a cell organelle that _____ cellular products.
9. The cell organelles that are the site of cell respiration are the _____.
10. The cell organelle that is a series of tunnels for transport within a cell is the _____.
11. One function of the endoplasmic reticulum of a cell is the synthesis of _____.
12. Lysosomes are cell organelles that contain _____ to destroy pathogens or damaged cellular parts.
13. The cell organelles that permit sperm cells to move are _____.
14. The cell organelles that sweep materials across a cell surface are _____.
15. The presence of _____ in a cell membrane will increase the surface area of the membrane.
16. Chromosomes are made of _____ and are found in the _____ of a cell.
17. In the process of facilitated diffusion, a substance moves from an area of _____ concentration to an area of _____ concentration.
18. The movement of molecules from an area of greater concentration to an area of lesser concentration is called _____.
19. The movement of molecules from an area of lesser concentration to an area of greater concentration is called _____.
20. The energy for the process of active transport is provided by _____.
21. The diffusion of water through a membrane is called _____.
22. A solution with a lesser salt concentration than human cells is said to be _____.
23. A solution with a greater salt concentration than human cells is said to be _____.
24. A solution with the same salt concentration as human cells is said to be _____.
25. Human cells will take in water and swell if they are placed in a solution that is _____.
26. Human cells will lose water and shrivel if they are placed in a solution that is _____.
27. The water content of red blood cells does not change when the cells are in blood plasma, because plasma is _____ to the cells.
28. White blood cells engulf bacteria by the process of _____.
29. Most human cells have _____ chromosomes, which is the _____ number for people.
30. The human haploid number of chromosomes is _____.

31. The genetic code of a chromosome is found in its _____.
32. In mitosis, the chromosomes become visible during _____.
33. In mitosis, the centrioles move to opposite poles of the cell during _____.
34. In mitosis, the pairs of chromatids line up on the equator of the cell during _____.
35. In mitosis, the chromatids become separate chromosomes during _____.
36. In mitosis, the spindle fibers contract and pull the sets of chromosomes during _____.
37. In mitosis, the two sets of chromosomes are separated during _____.
38. In mitosis, two nuclear membranes begin to form during _____.
39. After mitosis, the division of the cytoplasm is called _____.
40. The process of meiosis is necessary to form _____ cells and _____ cells.
41. The process of meiosis in the ovaries is called _____.
42. Spermatogenesis is the name for the process of _____, which takes place in the _____ to produce _____.
43. In mitosis, _____ division(s) produce(s) _____ cells, each with the _____ number of chromosomes.
44. In meiosis, _____ division(s) produce(s) _____ cells, each with the _____ number of chromosomes.
45. Within cells, the process of protein synthesis takes place on the organelles called _____.
46. In the process of protein synthesis, amino acids are brought to the mRNA by a molecule called _
47. The copying of the DNA genetic code by mRNA is called _____.
48. The lining up of amino acids according to the codons of an mRNA molecule to form a protein is called _____.

TISSUES
LAB 5

CRASHCOURSE VIDEO(S)

Click on the video embedded within your online platform or enter the address below into your web browser:
1. **https://youtu.be/i5tR3csCWYo**
2. https://youtu.be/lUe_RI_m-Vg
3. **https://youtu.be/D-SzmURNBH0**
4. **https://youtu.be/Jvtb0a2RXaY**

(Please watch the videos below before continuing)

DEFINING KEY TERMS:

1. Absorption:

2. Apical surface:

3. Basal surface:

4. Basement membrane:

5. Columnar:

6. Connective tissues:

7. Cuboidal:

8. Endocrine glands:

9. Epithelial tissues:

10. Exocrine glands:

11. Goblet cells:

12. Histology:

13. Intercellular junctions:

14. Muscle tissues:

15. Nervous tissues:

16. Secretion:

17. Simple:

18. Squamous:

19. Stratified:

20. Tissue:

Cells group together with one another based on similar structure and function to form tissues. Tissues provide the numerous functions of organs necessary to maintain biological life. Laboratory exercises in this section will seek to introduce students to the various tissues found in the human body and to familiarize them with their composition and function. The study of tissues is called histology, and is important to the understanding of how the human body is able to function as a unit.

EXERCISE 5.1
IDENTIFYING TISSUES UNDER THE MICROSCOPE

Purpose of exercise: To observe examples of the major tissue types under the microscope

Click on *Exercise 5.1 (a)* within your online platform or enter the address below into your web browser:
http://histologyguide.org/

Once you enter the site, click onto the **Slide Box** tab located on the left hand side. You will see a list of slide categorized by tissue type and organ system in bold font. Click onto the tabs that correctly identifies the tissue type you must observe for this exercise. The tissue type is listed on the table below.

If you encounter difficulties linking to this web address, or would like to view a different source try the link below.

Click on *Exercise 5.1 (b)* within your online platform or enter the address below into your web browser:
http://www.kumc.edu/instruction/medicine/anatomy/histoweb/index.htm

(Your instructor may choose to upload pictures of the tissue types in your college's online platform. If this is the case, you can still utilize the listed slides below as a reference to identify.)

Make a drawing of each prepared slide on high power in the circles below.

Prepared Slide *(Tissue Example)*	**Tissue Type**
1. Lung or Kidney	Simple squamous epithelium
2. Tubules of kidney	Simple cuboidal epithelium
3. Small intestine	Simple columnar epithelium
4. Trachea	Pseudostratified, ciliated, columnar epithelium

5.	Esophagus	Stratified squamous epithelium
6.	Urinary bladder	Transitional epithelium
7.	Areolar Connective tissue	Areolar connective tissue
8.	Adipose tissue	Adipose connective tissue
9.	Human skin, non-pigmented	Dense irregular connective tissue
10.	Dense regular Connective tissue	Dense regular connective tissue
11.	Hyaline Cartilage	Hyaline cartilage
12.	Epiglottis	Elastic cartilage
13.	Fibrocartilage	Fibrocartilage
14.	Compact bone, ground (human)	Bone
15.	Human blood smear	Blood
16.	Human smooth muscle	Smooth muscle
17.	Human skeletal muscle	Skeletal muscle
18.	Human cardiac muscle	Cardiac muscle
19.	Cerebrum	Nervous tissue

Observations

Tissue Type	Tissue Drawing on High Power (40X Objective)
1. Simple squamous epithelium	
2. Simple cuboidal epithelium	

3. Simple columnar epithelium	
4. Pseudostratified, ciliated, columnar epithelium	
5. Stratified squamous epithelium	
6. Transitional epithelium	

7. Areolar connective tissue	◯
8. Adipose tissue	◯
9. Dense irregular connective tissue	◯
10. Dense regular connective tissue	◯

11. Hyaline cartilage	
12. Elastic cartilage	
13. Fibrocartilage	
14. Bone	

15. Human blood smear

16. Human smooth muscle

17. Human skeletal muscle

18. Human cardiac muscle

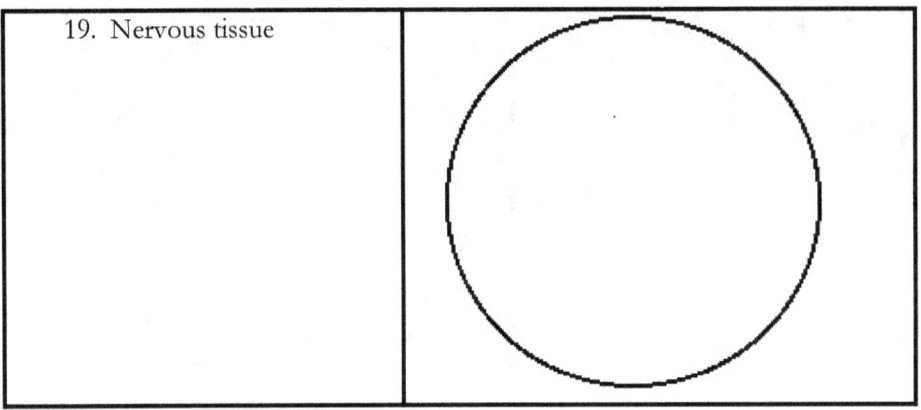

19. Nervous tissue

EXERCISE 5.2
CLASSIFICATION OF TISSUES

Purpose of exercise: To compare and contrast the four major tissue types.

Use the choices below to identify the major tissue types described. May have more than one correct answer.

Answer choices: a. connective tissue   b. epithelium   c. muscle   d. nervous tissue

1. _____ lines body cavities and covers the body's external surface
2. _____ transmits electrical signals
3. _____ anchors, packages, and supports body organs
4. _____ cells may absorb, secrete, and filter
5. _____ used to protect and provide structural support
6. _____ most involved in regulating and controlling body functions
7. _____ major function is to contract
8. _____ found making up the wall of blood vessels
9. _____ synthesizes hormones
10. _____ the most durable tissue type
11. _____ abundant nonliving extracellular matrix
12. _____ most widespread and abundant tissue type in the body
13. _____ forms the spinal cord and the brain

Answer the following questions.

1. Describe five general characteristics of epithelial tissue.

2. On what basis are epithelial tissues classified?

3. List five major functions of epithelium in the body, and give examples of each.

4. How does the function of stratified epithelia differ from the function of simple epithelia?

5. Where is ciliated epithelium found?

6. Transitional epithelium is actually stratified squamous epithelium with special characteristics. How does it differ structurally from other stratified squamous epithelia?

7. How do the endocrine and exocrine glands differ in structure and function?

8. What are three general characteristics of connective tissues?

9. What functions are performed by connective tissue?

10. The three types of muscle tissue exhibit similarities as well as differences. Check the appropriate space in the chart to indicate which muscle types exhibit each characteristic.

Characteristic	Skeletal	Cardiac	Smooth
Voluntarily controlled			
Involuntarily controlled			
Striated			
Has a single nucleus in each cell			
Has several nuclei per cell			
Found attached to bones			
Allows you to direct your eyeballs			
Found in the walls of the stomach, uterus, and arteries			
Contains spindle-shaped cells			
Contains branching cylindrical cells			
Contains long, nonbranching cylindrical cells			
Has intercalated discs			
Concerned with locomotion of the body as a whole			
Changes the internal volume of an organ as it contracts			
Tissue of the heart			

EXERCISE 5.3
REVIEW QUESTIONS

1. Diffusion of gases in the lungs is possible because the alveoli are made of _____ epithelium.
2. The tissue that removes dust and pathogens from the trachea is _____.
3. The tissue that keeps the trachea an open airway is _____.
4. The epidermis of the skin is made of _____ epithelium to help prevent water loss.
5. The urinary bladder is lined with _____, in which the cells stretch and flatten.
6. Organs that contain epithelial tissue capable of secretion are called _____.
7. The tissue that protects the brain from mechanical injury is _____.
8. The tissue that forms smooth surfaces in joints is _____.
9. Ligaments are made of _____ tissue.
10. The most important physical characteristic of the protein collagen is that it is _____.
11. Chondrocytes are the cells found in _____.
12. The most important physical characteristic of the protein elastin is that it is _____.
13. Collagen and elastin are _____ molecules made by cells called _____.
14. The function of red blood cells is to _____.
15. The protein hemoglobin is found in _____ cells and contains the mineral
16. The blood cells that destroy pathogens are the _____ cells.
17. The function of platelets is _____.
18. The part of a neuron that carries impulses toward the cell body is the _____.
19. The part of a neuron that carries impulses away from the cell body is the _____.
20. The space between two nerve cells is a _____.
21. Electrical insulation for nerve cells is provided by the _____.
22. The heart is made of _____, which _____.
23. The _____ tissue of the small intestine creates waves of contraction called peristalsis.
24. The specific tissue that produces a large amount of body heat is _____.
25. The type of muscle tissue found in arteries and veins is _____.
26. Glands that have ducts are called _____ glands.
27. The tissue that stores potential energy in the form of true fats is _____.

THE INTEGUMENTARY SYSTEM
LAB 6

CRASHCOURSE VIDEO(S):

Click on the video embedded within your online platform or enter the address below into your web browser:
1. **https://youtu.be/Orumw-PyNjw**
2. **https://youtu.be/EN-x-zXXVwQ**

(Please watch the videos below before continuing)

DEFINING KEY TERMS:

1. Apocrine glands:

2. Arrector Pili Muscle:

3. Dermis:

4. Eccrine (Merocrine) Glands:

5. Epidermis:

6. Keratinization:

7. Lamellated (Pacinian) corpuscles:

8. Lunula:

9. Melanin:

10. Papillary Layer:

11. Reticular Layer:

12. Sebaceous Glands:

13. Sebum:

14. Stratum Basale:

15. Stratum Corneum:

16. Stratum Granulosum:

17. Stratum Lucidum:

18. Stratum Spinosum:

19. Subcutaneous (hypodermis):

20. Tactile (Meissner's) corpuscles:

EXERCISE 6.1
IDENTIFYING INTEGUMENTARY TISSUES UNDER THE MICROSCOPE

Purpose of exercise: To observe examples of integumentary tissues under the microscope.

Click on *Exercise 6.1 (a)* within your online platform or enter the address below into your web browser:
http://histologyguide.org/

Once you enter the site, click onto the **Slide Box** tab located on the left hand side. You will see a list of slide categorized by tissue type and organ system in bold font. Click onto the tabs that correctly identifies the tissue type you must observe for this exercise. The tissue type is listed on the table below.

If you encounter difficulties linking to this web address, or would like to view a different source try the link below.

Click on *Exercise 6.1 (b)* within your online platform or enter the address below into your web browser:
http://www.kumc.edu/instruction/medicine/anatomy/histoweb/index.htm

(Your instructor may choose to upload pictures of the tissue types in your college's online platform. If this is the case, you can still utilize the listed slides below as a reference to identify.)

Make a drawing of each prepared slide on high power in the circles below.

Prepared Slide (*Tissue Example*)
1. Human Heavily Pigmented Skin
2. Human Nonpigmented Skin
3. Plantar Skin
4. Human Skin showing Sweat Glands
5. Human Scalp
6. Human Hair

Prepared Slide *(Tissue Example)*	Tissue Drawing on High Power (40X Objective)
1. Human Heavily Pigmented Skin	
2. Human Nonpigmented Skin	
3. Plantar Skin	
4. Human Skin showing Sweat Glands	

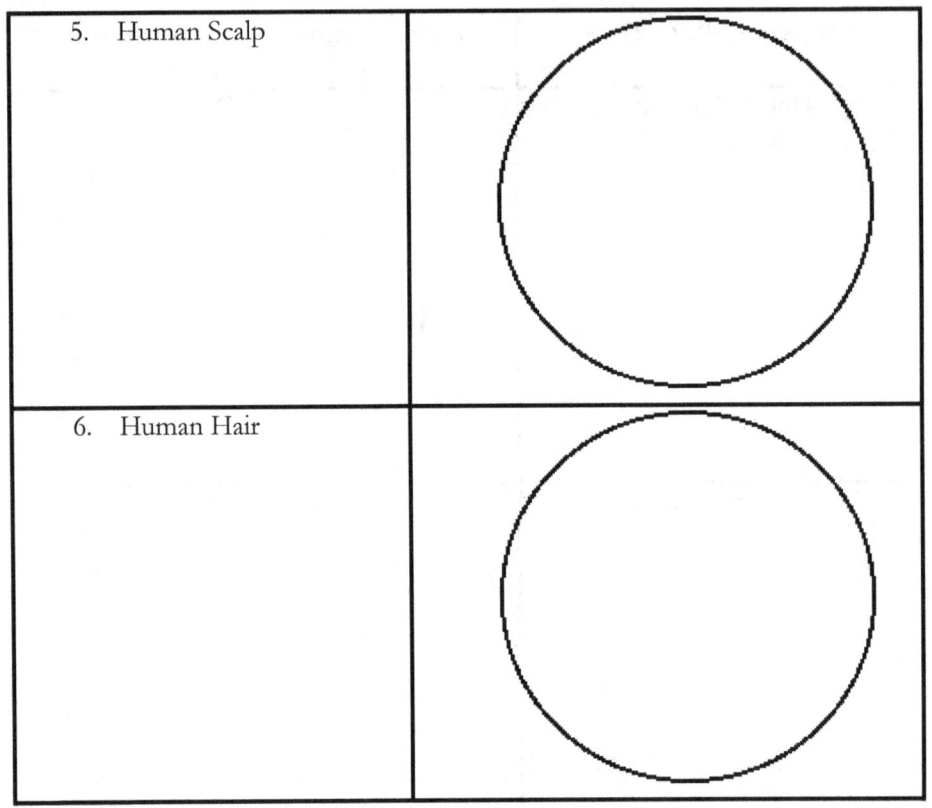

5. Human Scalp	
6. Human Hair	

EXERCISE 6.2
LABEL THE DIAGRAM OF THE SKIN

Purpose of exercise: To identify parts of the human integumentary system.

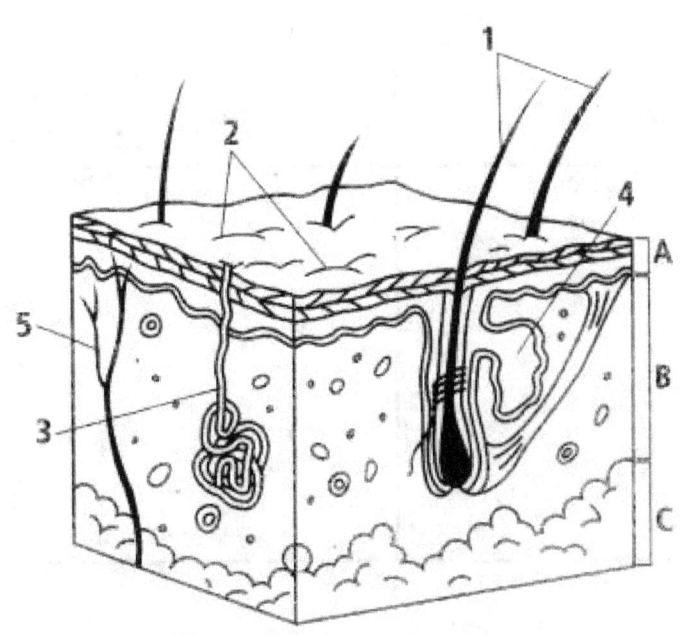

EXERCISE 6.3
CHARACTERISTICS OF THE SKIN

Purpose of exercise: To observe the function of the skin and its accessory structures.

Materials:
- Clear drinking glass (*with a clear bottom*)
- Coin
- Millimeter ruler
- Color pencils
- Two different color markers
- A lab partner

Procedure

Feature of the Fingernails

1. Examine your (unpainted, natural) fingernail.
2. Draw what you see in the circle below.

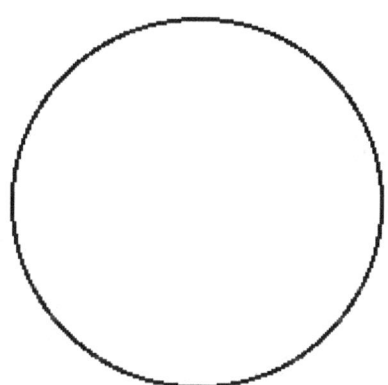

Visualizing Changes in Skin Color Due to Continuous External Pressure

1. Obtain a Clear drinking glass.
2. Press the heal of your hand firmly against the bottom of the drinking glass for a few seconds and then observe the color of your skin in the compressed area by looking through the glass.

Questions:
1. What is the color of the compressed skin?

2. Why does the color of the skin change?

3. What would happen if the pressure was continued for an extended period in this area?

Testing Tactile Localization

Your skin has many sensory "touch" receptors. Tactile localization is the ability to determine which portion of the skin has been touched. Once the skin's sensory receptors have received a message it sends this message to the brain and then the brain interprets the location and "meaning" of the feeling. (Rough, Smooth, Soft, Tickly, Painful, etc.) The more sensory receptors in an area of the skin, the more accurately the brain can interpret the location.

1. Make sure your lab partner's eyes are closed. The experimenter will now touch the palm of the subject's hand with a marker. The subject should then try to touch the exact point with his/her own different color marker.
   a) What happened?
2. Using a ruler, measure the error of localization in millimeters (the distance between the 2 marks).
3. Repeat the test in the same spot two more times, recording the error of localization for each test.
4. Average the results of the three trials and record your data in the chart.
5. Repeat this procedure on the fingertip, ventral forearm, and the back of hand.

Testing Tactile Location Data Table

Body Area Tested	Average error of localization
Palm of hand	_____
Fingertip	_____
Ventral forearm	_____
Back of hand	_____

Questions:
1. Does the ability to localize the stimulus improve the second time? The third time?

2. Why do you think this happened?

3. Which area has the smallest error of localization and is therefore the most sensitive to touch?

Demonstrating Adaptation of Touch Receptors

In many cases, when a stimulus is applied for a long period of time, the rate at which the receptors respond slows down. Your awareness of the stimulus declines or is lost until a change in the stimulus occurs. This is referred to as ADAPTATION. The touch receptors adapt.

1. Make sure your subject's eyes are closed.
2. Place a coin on the anterior surface of the subject's forearm, and determine how long the sensation persists for the subject. Using a timer, record the duration of the sensation in seconds in the chart.
3. Repeat the test, placing the coin at a different forearm location. Record the duration of the sensation.
4. After awareness of the sensation has been lost at the second site, stack three more coins atop the first one. Ask, does the pressure sensation return? If so, record the duration of the sensation.

Question:
Are the same receptors being stimulated when the four coins, rather than one coin, are used? Explain your reasoning.

EXERCISE 6.4
SKIN COLOR

Purpose of exercise: To summarize the factors that determine skin color.

Click on *Exercise 6.4* within your online platform or enter the address below into your web browser:
http://media.hhmi.org/biointeractive/interactivevideo/skincolorquiz/?_ga=2.71526355.1125362674.1506365954-818520368.1502809801

If you have trouble, try this hyperlink:
https://youtu.be/hFw8mMzH5YA

Please watch the video

EXERCISE 6.5
PLOTTING THE DISTRIBUTION OF SWEAT GLANDS IN THICK VS. THIN SKIN

Purpose of exercise: To assess sweat glands of the skin

In this experiment, you will test two parts of your body (your forearm and your palm) for sweat gland activity. To do this, you will paint a small amount of iodine in both locations. Once dry, you will tape a small paper square on your skin. As you sweat, the iodine will be transferred from your skin to the paper, turning the starch in the paper blue-black. The amount of blue-black on the two papers will be used to conclude the relative numbers of active sweat glands in the two areas.

Materials
- Iodine
- Two Small squares (1cm x 1cm) of white copy paper that have dried cornstarch solution on them
- Roll of adhesive tape
- Cotton swab

Procedure
1. Using a cotton swab, each member of the lab group should paint a small amount of iodine on both the medial aspect of the palm and on the forearm of your non-writing arm.
2. Once the iodine is dry, use the tape to secure a small piece of paper over each dried iodine spot. Tape the paper directly to your skin.
3. Leave the paper in contact with your skin for 20 min. During this time, you can do other parts of the lab.
4. When the 20 min has passed, peel the paper off your skin and determine whether or not a significant portion of the paper has turned blue-black or has remained white. Blue-black color is positive for sweating; white is negative for sweating

Questions:
1. Which of the two areas of your body that you tested contained more sweat glands?
2. Why do we have more sweat glands in this area than in the other area?

EXERCISE 6.6
FINGERPRINT CHARACTERISTICS

Purpose of exercise: To distinguish and identify features of the human fingerprint.

Skin prints are a way to take a closer look at your cutaneous membrane.

Materials
- Pencil
- Clear Tape

Procedure
1. Rub soft pencil lead shavings onto a sheet of scratch paper until you get a black blob.
2. Pick up a good smudge on your finger.
3. Carefully pick up the smudge with a piece of clear tape.
4. Pull it away and tape it on to your lab paper.
5. Use the observation table to record the skin prints of each finger on the hand that you do not write with.
6. Examine the fingerprints carefully and compare them to the fingerprint examples below.

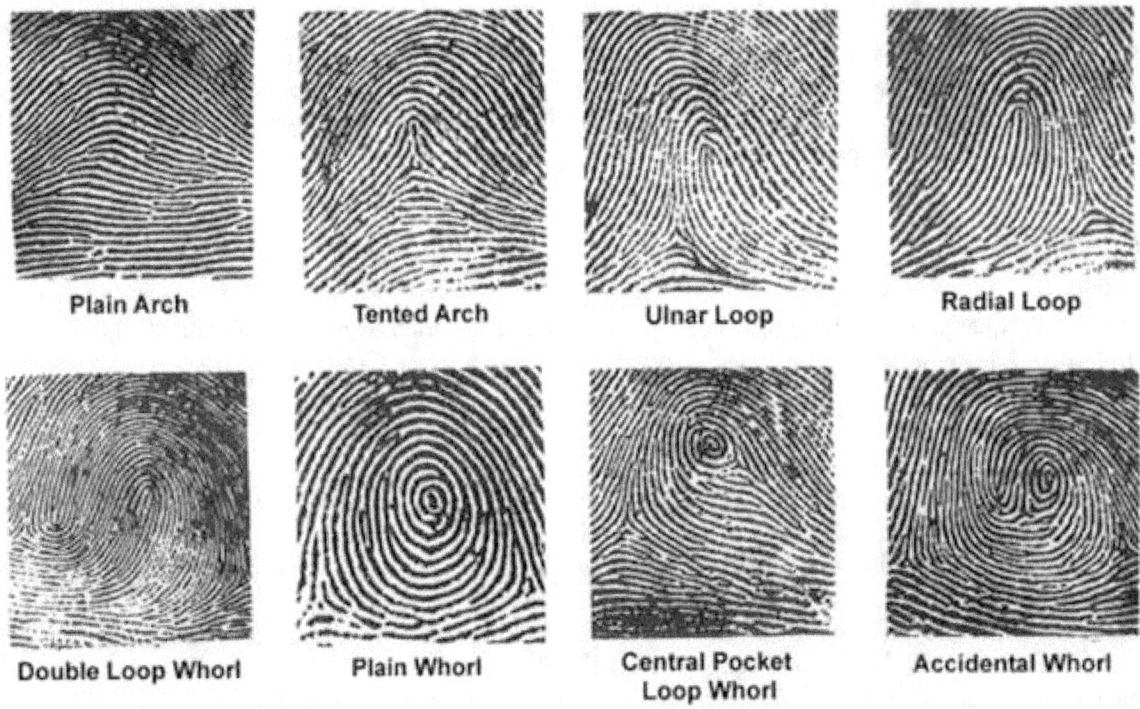

7. Identify the loop pattern on each fingerprint and write it beside the fingerprint. Identify each finger using: thumb, index finger, middle finger, ring finger, pinky finger
8. Did all of your fingers have the same loop pattern?

Observation Table

Name of Finger	Fingerprint *(Place your Fingerprint here)*	Name of your Fingerprint Type
1. Thumb		
2. Index Finger		
3. Middle Finger		
4. Ring Finger		
5. Pinky Finger		

EXERCISE 6.7
FURTHER YOUR UNDERSTANDING

1. Fill in the blanks of the following sentence using the wordlist provided below.

| Dermis   Dermatology   Hypodermis   epidermis   nails   Glands |

The integumentary system consists of the skin, and its accessory organs (_____, _____, and _____). The skin has two major layers: _____ & _____. The subcutaneous region of the skin is called _____. The study of the integumentary system is called _____.

2. List the general functions of the skin and subcutaneous layer.

3. What type of tissue is the epidermis?
4. In which layer of the epidermis are cells dividing?
5. Where on your body would you expect to find cornified skin?
6. Which layer contains melanocytes?
7. What is the name of the pigment that they produce? What purpose does it serve?
8. Which layer(s) show pigmentation?

EXERCISE 6.8
SURGERIES OF THE INTEGUMENTARY SYSTEM

Purpose of exercise: To explore various types of surgeries associated with the human integumentary system.

***Please make sure that your Adobe Flash Player is updated on your computer. Also, please be care not to click on any of the third-party advertisements because they will route you to another site.

You will perform the following virtual surgeries: **Laser Tattoo Removal, Hair Transplant Procedure, Laser Hair Removal, Ingrown Toenail Removal,** and **Tumescent Liposuction.**

Enter the following addresses for each surgery type into your web browser. Once you are ready to begin the surgery, click **START**. Please listen and read the instructions provided on the website and follow the interactive steps to perform the various surgeries.

Click on *Exercise 6.8 (a)* within your online platform or enter the address below into your web browser.
Laser Tattoo Removal: http://www.surgerysquad.com/surgeries/virtual-laser-tattoo-removal/

Click on *Exercise 6.8 (b)* within your online platform or enter the address below into your web browser.
Hair Transplant Procedure: http://www.surgerysquad.com/surgeries/virtual-hair-transplant-surgery/

Click on *Exercise 6.8 (c)* within your online platform or enter the address below into your web browser.
Laser Hair Removal: http://www.surgerysquad.com/surgeries/virtual-laser-hair-removal/

Click on *Exercise 6.8 (d)* within your online platform or enter the address below into your web browser.
Ingrown Toenail Removal: http://www.surgerysquad.com/surgeries/ingrown-toenail-removal/

Click on *Exercise 6.8 (e)* within your online platform or enter the address below into your web browser.
Tumescent Liposuction: http://www.surgerysquad.com/surgeries/tumescent-belly-liposuction-surgery/

EXERCISE 6.9
SKIN CANCER INVESTIGATION

Purpose of exercise: To learn about healthy skin and how to protect it from the damaging effects of UV rays from the sun. Also, to learn about three types of skin cancer and how to identify them.

Click on *Exercise 6.9* within your online platform or enter the address below into your web browser:
http://www.sciencenetlinks.com/skindeep/interactive/base.html

Move through the activity by clicking on all of the suggested links.

Healthy Skin
1. What are the types of cells in the epidermis and how are they related to each other?

2. What is melanin and what is its role in the epidermis?

3. What is keratin and what does it do?

4. What two proteins does the dermis contain? What qualities do these give the dermis?

5. What seven other structures does the dermis contain?

6. Describe the subcutaneous tissue.

Effects of the Sun on the Skin
7. What are some of the benefits of the sun's UV rays?

8. How are these UV rays harmful?

Skin Cancer: Causes
9. Explain how UV rays can affect normal cell division and potentially lead to cancer.

Skin Cancer: Benign and Malignant Tumors
10. What are the five characteristics of malignant tumors?

Skin Cancer: Common Risk Factors
11. Describe each of the five risk factors:

   Actinic keratosis -

   Genetics -

   Environment -

   Complexion -

   Age -

Skin Cancer: Types
12. What is the primary symptom of any skin cancer?

13. Complete the following table of the three types of skin cancer:

	Basal Cell Carcinoma	Squamous Cell Carcinoma	Melanoma
Definition			
Characteristics			
Risk Factors			
Symptoms			

Skin Cancer: Warning Signs of Malignant Melanoma

14. Summarize the four warning signs of malignant melanoma.

Asymmetry -

Borders -

Color -

Diameter -

Skin Cancer: Prevention and Detection
15. Describe the characteristics of proper sunscreen application.

16. When is sunlight most intense?

Visit the Glowell Clinic and move through all of the activities.

The Lab
Fill in the table for each of the patient files:

Patient	Age/Race/Location of Growth	Checklist of Symptoms	Cancerous or Non-cancerous?
A			
B			
C			
D			
E			
F			

The Helpdesk

Summarize what advice you give to each of the callers.

Caller 1 -

Caller 2 -

Caller 3 -

Caller 4 -

Caller 5 -

Caller 6 -

EXERCISE 6.10
REVIEW QUESTIONS

1. The outer layer of the skin is the _____, which is made of _____ tissue.
2. The inner layer of the skin is the _____, which is made of _____ tissue.
3. The lowest layer of the _____ is the stratum germinativum, and the function of its cells is _____.
4. Keratin is a _____ (type of molecule) that helps make the skin relatively waterproof.
5. The outermost layer of the epidermis is the _____.
6. Melanocytes produce the protein _____ when stimulated by _____.
7. The structure in which a hair grows is called a _____, and the process by which a hair grows is _____.
8. The ends of fingers and toes are protected from mechanical injury by _____.
9. Drying of the skin is prevented by _____.
10. The glands that secrete sebum are called _____.
11. Drying of the eardrum is prevented by _____.
12. The purpose of sweating is to increase the loss of _____, but it has the disadvantage of leading to _____.
13. In a cold environment, the arterioles in the dermis will _____.
14. In a warm environment, the arterioles in the dermis will _____.
15. Vitamin _____ is produced in skin that is exposed to _____.
16. Between the dermis and the muscles is the _____.
17. The layer of the integumentary system that provides some insulation from cold is the _____.

SKELETAL SYSTEM
LAB 7

CRASHCOURSE VIDEO(S)

Click on the video embedded within your online platform or enter the address below into your web browser
1. **https://youtu.be/rDGqkMHPDqE**
2. **https://youtu.be/DLxYDoN634c**

(Please watch the videos below before continuing)

DEFINING KEY TERMS:

1. Appendicular Skeleton:

2. Articular Cartilage:

3. Axial Skeleton:

4. Calcitonin:

5. Chondrocyte:

6. Collagen:

7. Diaphysis:

8. Endochondral Ossification:

9. Endosteum:

10. Epiphyseal Line:

11. Epiphyses:

12. Foramen:

13. Hematopoietic Tissue:

14. Intramembranous Ossification:

15. Ligament:

16. Medullary Cavity:

17. Metaphysis:

18. Osteoblasts:

19. Osteoclasts:

20. Osteocytes:

21. Osteon:

22. Paranasal Sinuses:

23. Parathyroid hormone (PTH):

24. Periosteum:

25. Red Bone Marrow:

26. Spongy Bone:

27. Sutures:

28. Yellow Bone Marrow:

EXERCISE 7.1
IDENTIFYING OSSEOUS TISSUES UNDER THE MICROSCOPE

Purpose of exercise: To observe examples of osseous tissues under the microscope.

Click on *Exercise 7.1* within your online platform or enter the address below into your web browser:
http://histologyguide.org/

Once you enter the site, click onto the **Slide Box** tab located on the left hand side. You will see a list of slide categorized by tissue type and organ system in bold font. Click onto the tabs that correctly identifies the tissue type you must observe for this exercise. The tissue type is listed on the table below.

If your encounter difficulties linking to this website, or would like to view different slides try this link instead:
http://www.kumc.edu/instruction/medicine/anatomy/histoweb/index.htm

(Your instructor may choose to upload pictures of the tissue types in your college's online platform. If this is the case, you can still utilize the listed slides below as a reference to identify.)

Prepared Slide *(Tissue Example)*	
1.	Compact Bone
2.	Developing Membrane Bone
3.	Human Spongy Bone
4.	Bone Marrow

Prepared Slide *(Tissue Example)*	Tissue Drawing on High Power (40X Objective)
1. Compact Bone	
2. Developing Membrane Bone	
3. Human Spongy Bone	
4. Bone Marrow	

EXERCISE 7.2
OSSEOUS TISSUE QUESTIONS

Please answer the questions in your laboratory manual.

1. The type of bone that is made of osteons is _____ bone.
2. New bone matrix for growth is produced by cells called _____.
3. Reabsorption of bone matrix is the function of the cells called _____.
4. Red bone marrow produces _____, _____, and _____.
5. The type of bone marrow that is found in the diaphysis of long bones is _____ marrow, which is mostly _____ tissue.
6. Compression of a baby's head during birth is permitted by the presence of _____ between cranial bones.
7. The embryonic humerus and femur are both made of _____.
8. In a child's long bones, growth occurs at the _____.
9. Long bones stop growing when all of their _____ has been replaced by _____.
10. The inherited maximum height a child can attain is called the _____.
11. Calcium and phosphorus are nutrients necessary to become part of the bone _____.
12. The hormone in women that promotes closure of the epiphyseal discs is _____.
13. The hormone in men that promotes closure of the epiphyseal discs is _____.

EXERCISE 7.3
PARTS OF THE HUMAN SKELETON

Purpose of exercise: To identify parts of the articulated human skeleton.

Click on *Exercise 7.4* within your online platform or enter the address below into your web browser:
http://www.eskeletons.org/boneviewer/nid/12537/region/skull/bone/cranium

Familiarize yourself with the virtual skeleton at this link. Choose and view different parts of the skeleton by clicking on the skeleton model located on the left of your screen. The link you've been provided will start you off looking at the human skull. If you get lost, you can return to the www.eskeletons.org home page and look under the **TAXON: HUMAN male, adult**. Once you are comfortable with your understanding of the human skull, move on to observing other parts of the axial and appendicular skeleton. Then complete the skeleton labeling exercises in your laboratory manual.

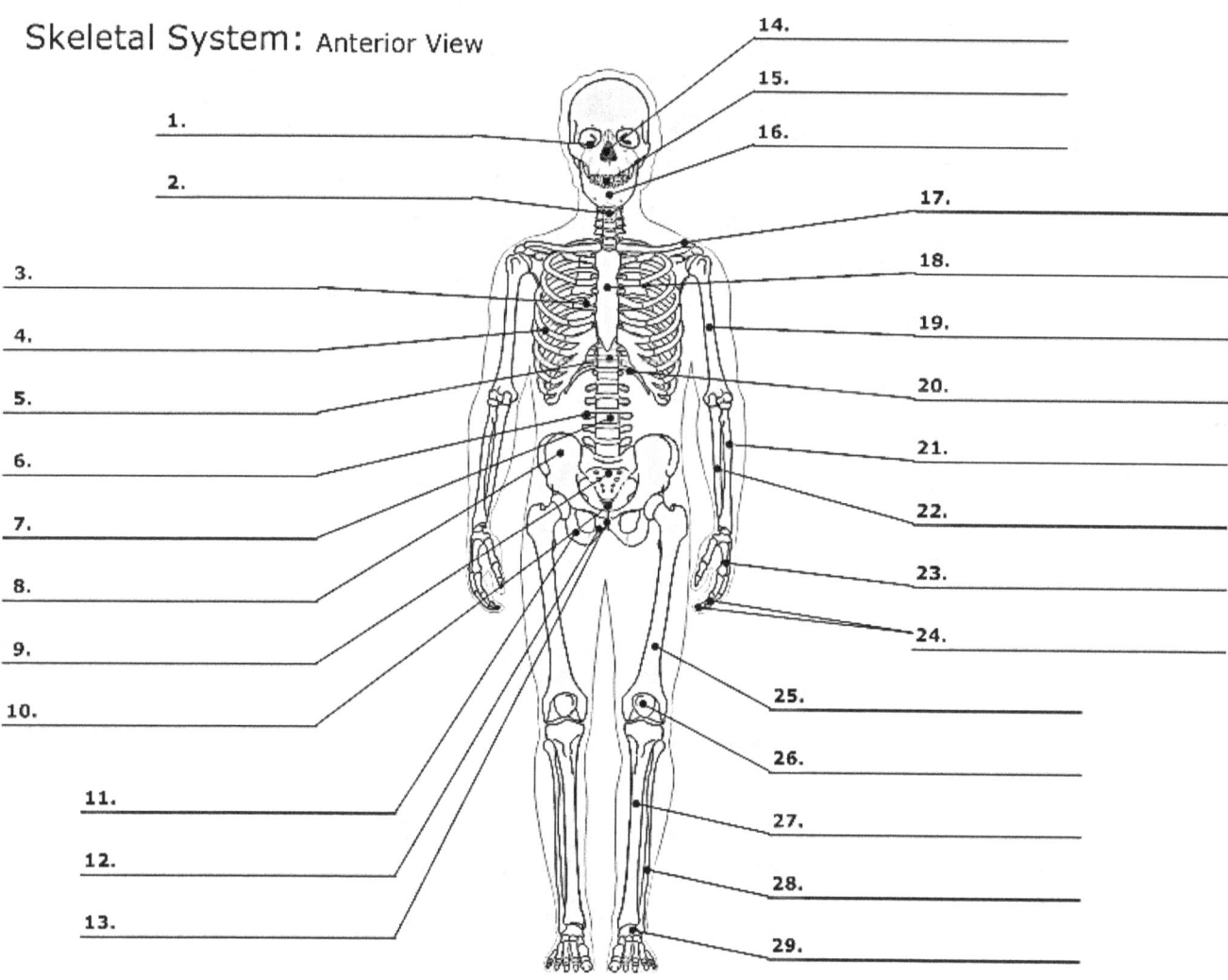

Skeletal System: Posterior View

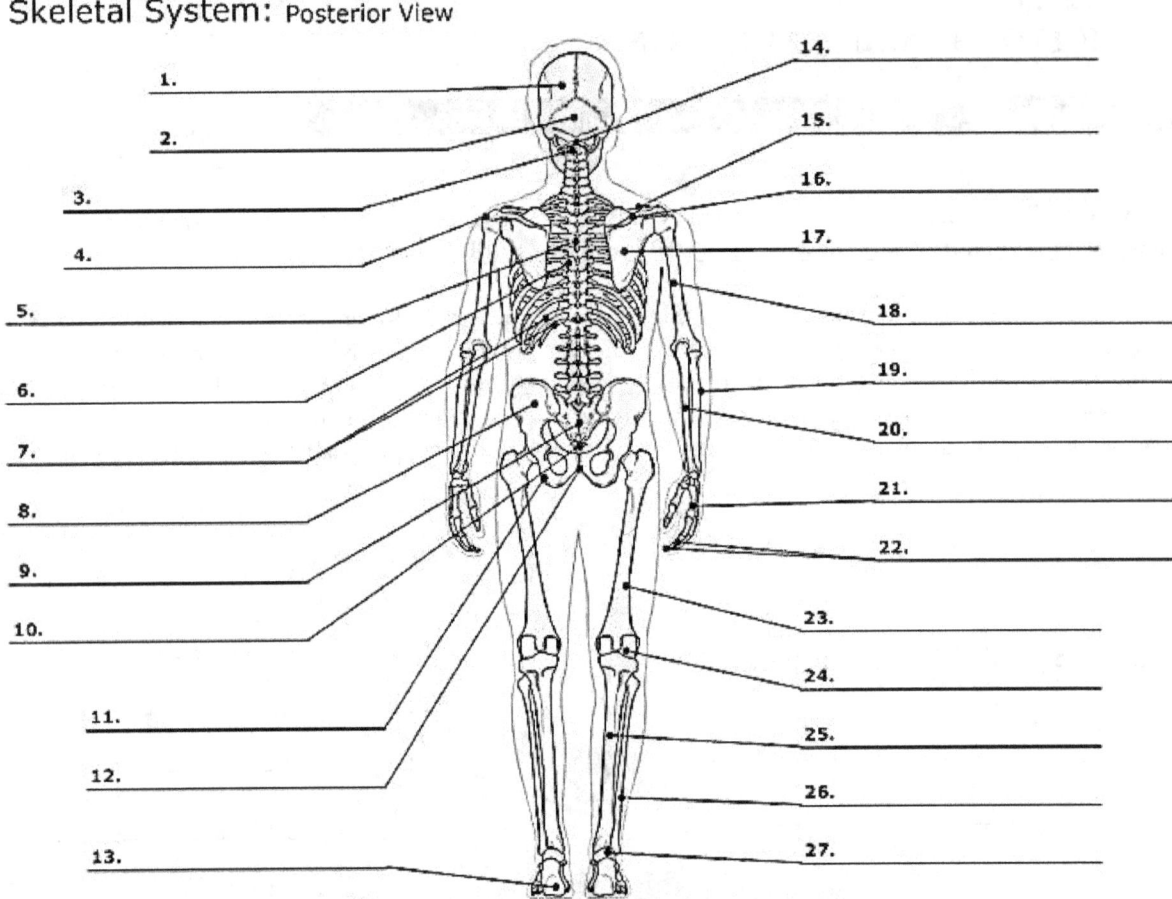

1. _____
2. _____
3. _____
4. _____
5. _____
6. _____
7. _____
8. _____
9. _____
10. _____
11. _____
12. _____
13. _____
14. _____
15. _____
16. _____
17. _____
18. _____
19. _____
20. _____
21. _____
22. _____
23. _____
24. _____
25. _____
26. _____
27. _____

EXERCISE 7.4
AXIAL SKELETON QUESTIONS

Please answer the questions in your laboratory manual.

1. List the 3 major portions of the axial skeleton.

2. Name the two sets of bones that make up the skull.

3. Tell what each of the two sets of bones in the skull accomplishes.

4. How many bones are in the cranium?

5. Tell the significance of each of the following: external auditory meatus, styloid process, zygomatic process, mastoid process, jugular foramen.

6. How many bones compose the face?

7. Name the functions of the paranasal sinuses.

8. What material makes up the tip of the nose?

9. What bone forms the septum in your nose?

10. What is the function of the hyoid bone?

11. List three ways that an infant's skull is different from that of an adult.

12. What are fontanels?

13. What happens to the fontanels as the baby grows?

14. How many bones are in the vertebral column?

15. How many bones are in each of the following sections of the spine?

16. What separates the vertebrae?

17. What happens to the water content of these discs with age?

18. Define each of these conditions: scoliosis, kyphosis, and lordosis.

19. Give the names of the C_1 and C_2 vertebrae.

20. What is the significance of the thoracic vertebrae?

21. Which type of vertebra(e) have transverse processes that have facets for articulation with the ribs?

22. How many ribs make up the thoracic cage?

23. Which ribs are true ribs? Which ribs are false ribs? Which ribs are floating ribs?

24. What is the space between ribs called?

EXERCISE 7.5
LABELING PARTS OF THE AXIAL SKELETON

Purpose of exercise: To identify parts of the human axial skeleton.

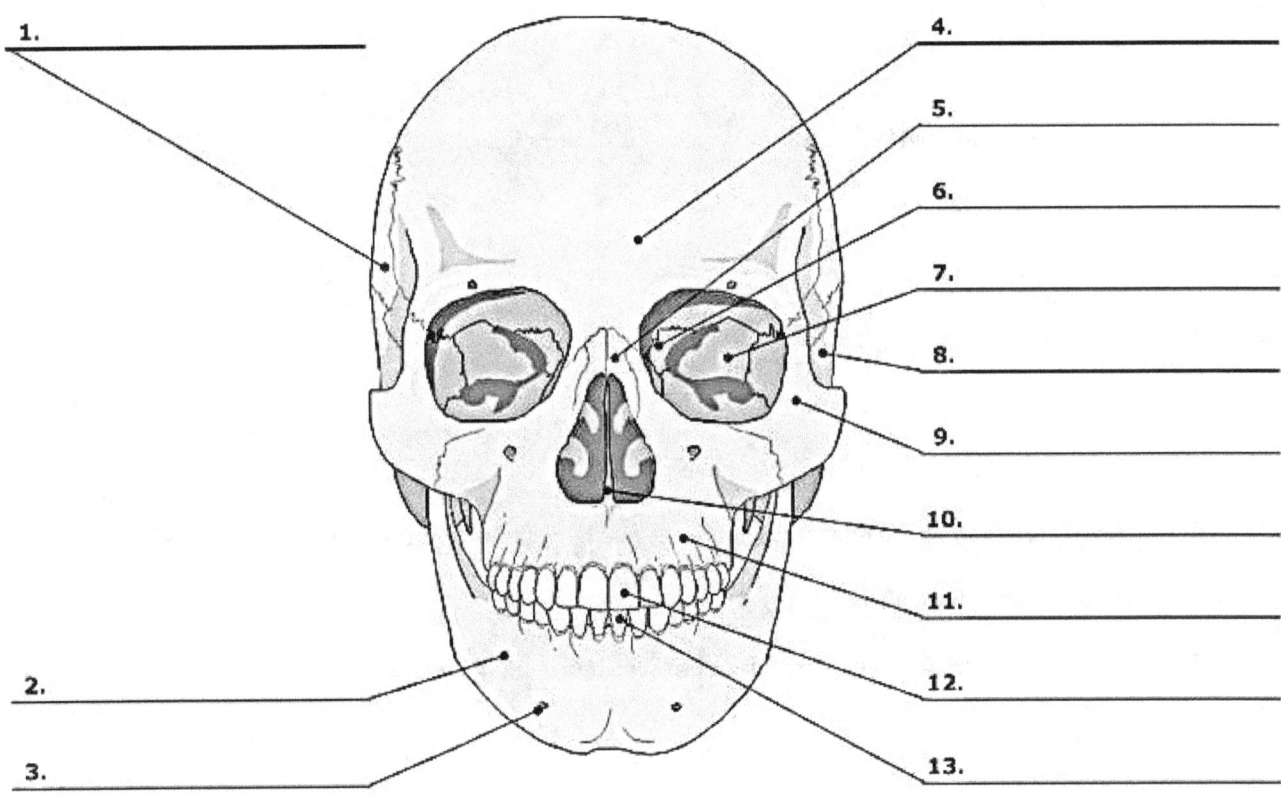

Bones of the Skull
Anterior View

1. _____
2. _____
3. _____
4. _____
5. _____
6. _____
7. _____
8. _____
9. _____
10. _____
11. _____
12. _____
13. _____

110

Bones of the Skull
Lateral View

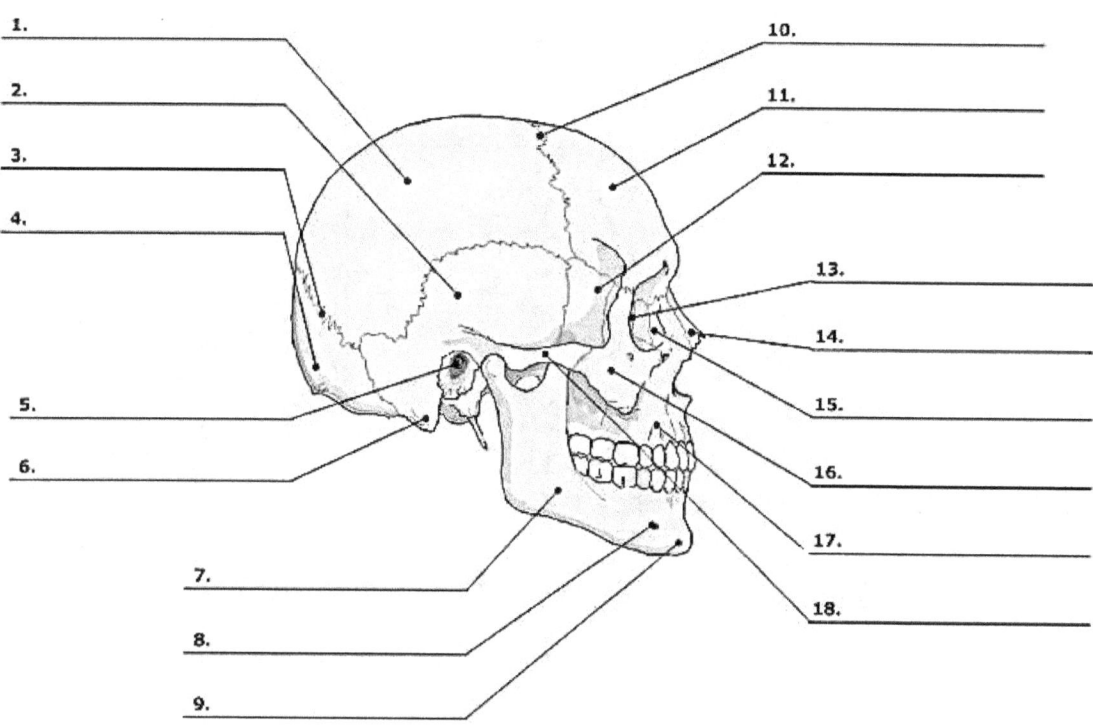

1. _____
2. _____
3. _____
4. _____
5. _____
6. _____
7. _____
8. _____
9. _____
10. _____
11. _____
12. _____
13. _____
14. _____
15. _____
16. _____
17. _____
18. _____

Bones of the Skull
Inferior View

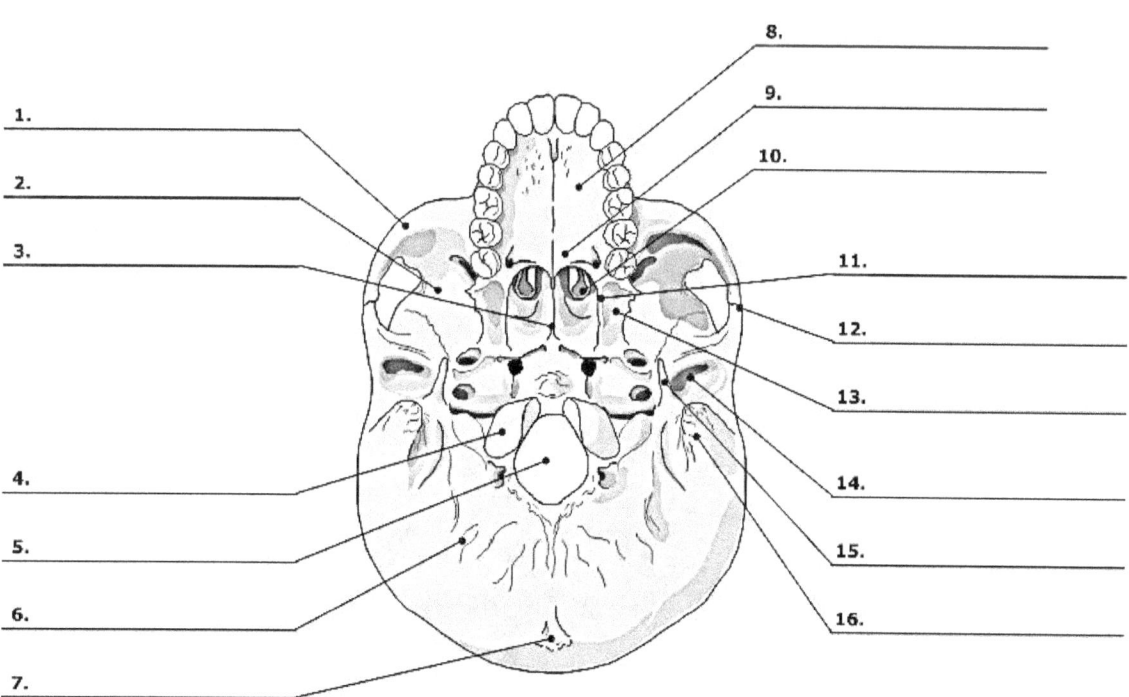

1. _____
2. _____
3. _____
4. _____
5. _____
6. _____
7. _____
8. _____
9. _____
10. _____
11. _____
12. _____
13. _____
14. _____
15. _____
16. _____

Rib Cage

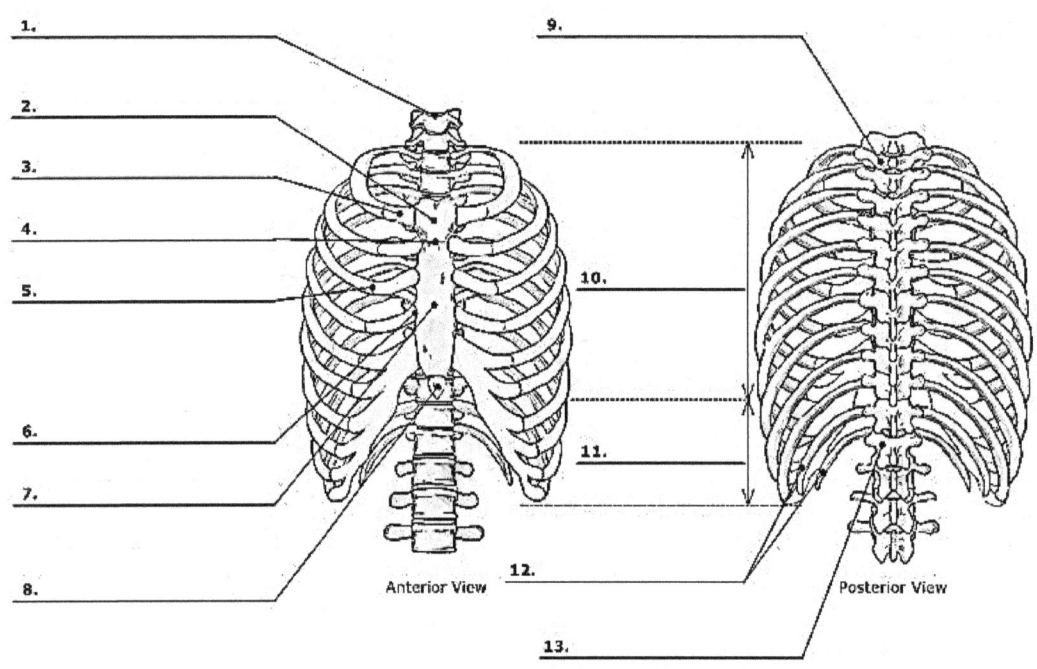

1.
2.
3.
4.
5.
6.
7.
8.
9.
10.
11.
12.
13.

Anterior View Posterior View

Vertebral Column

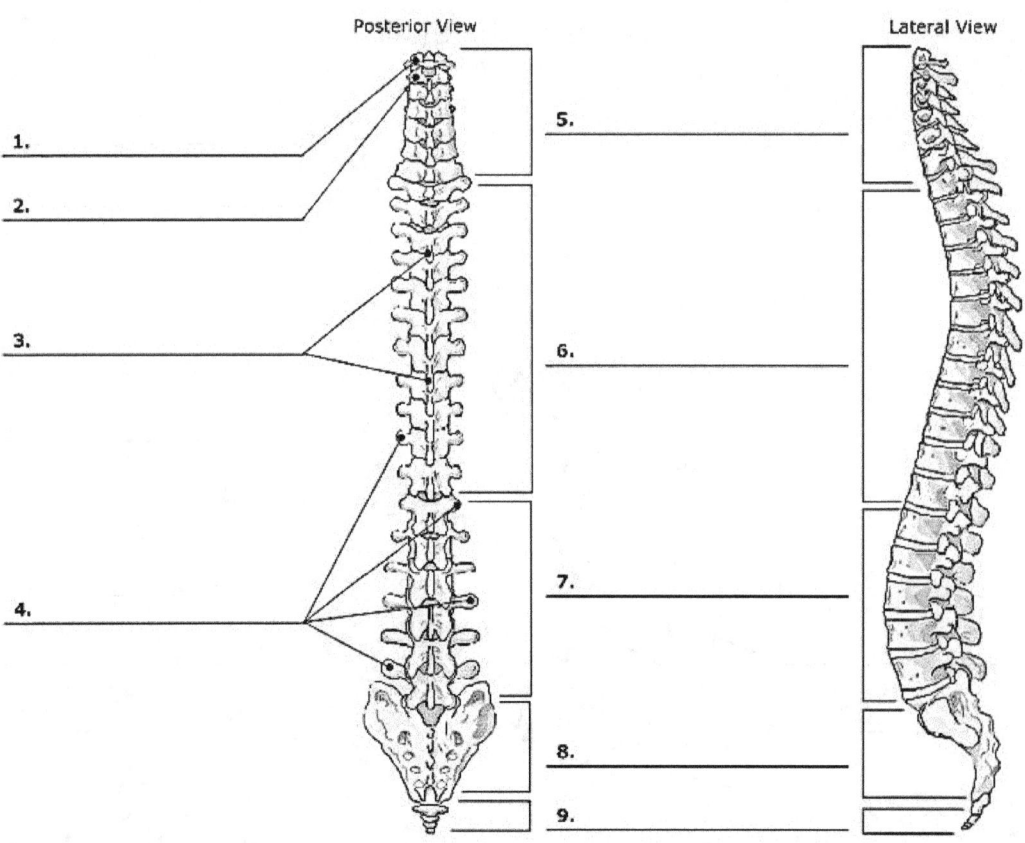

Posterior View   Lateral View

1.
2.
3.
4.
5.
6.
7.
8.
9.

Vertebral Column: Vertebra

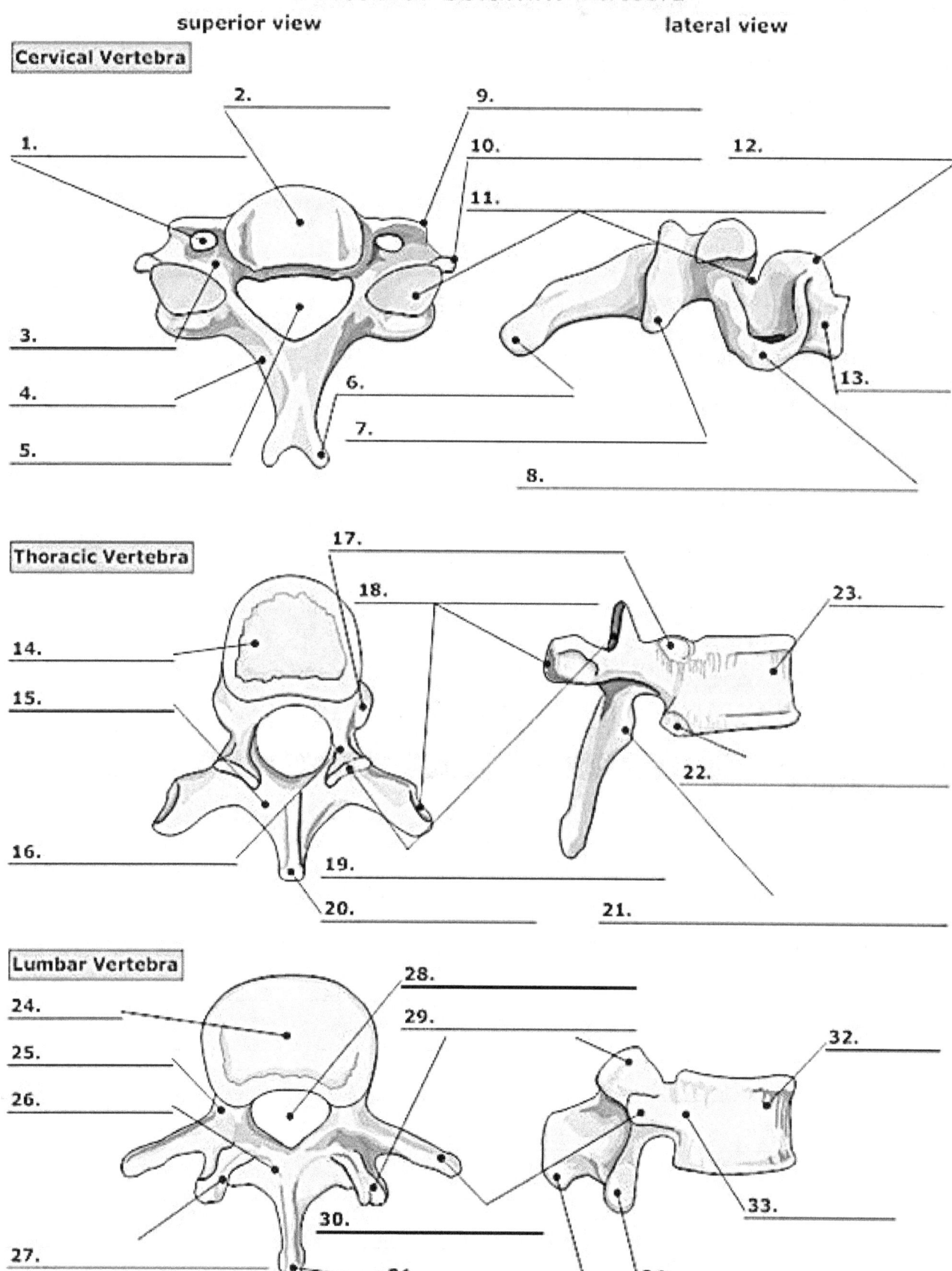

EXERCISE 7.6
APPENDICULAR SKELETON QUESTIONS

Please answer the questions in your laboratory manual.

1. How many bones make up the appendicular skeleton?
2. What is the name of the bone typically called the arm?
3. What are the names of the two bones of the forearm?
4. Name the three groups of bones that make up the hand.
5. What is the most important function of the pelvic girdle?
6. What is the name of the joint where the two pubic bones are attached?
7. What is then name of the socket for the thigh?
8. What is the name of the bone commonly called the thigh bone?
9. Which bone forms the ankle?
10. Name the seven tarsal bones.
11. Name the eight carpal bones.

EXERCISE 7.7
CARPAL AND TARSAL BONES MNEMONICS

Purpose of exercise: To create a mnemonic to assist in the identification of the eight carpal and seven tarsal bones.

A mnemonic is an instructional strategy designed to help students improve their memory of important information. This technique connects new learning to prior knowledge through the use of visual and/or acoustic cues. The basic types of mnemonic strategies rely on the use of key words, rhyming words, or acronyms.

In this exercise, you will come up with two mnemonics of your own for the carpal and tarsal bones.

_____

_____

_____

_____

_____

EXERCISE 7.8
LABELING PARTS OF THE APPENDICULAR SKELETON

Purpose of exercise: To identify parts of the human appendicular skeleton.

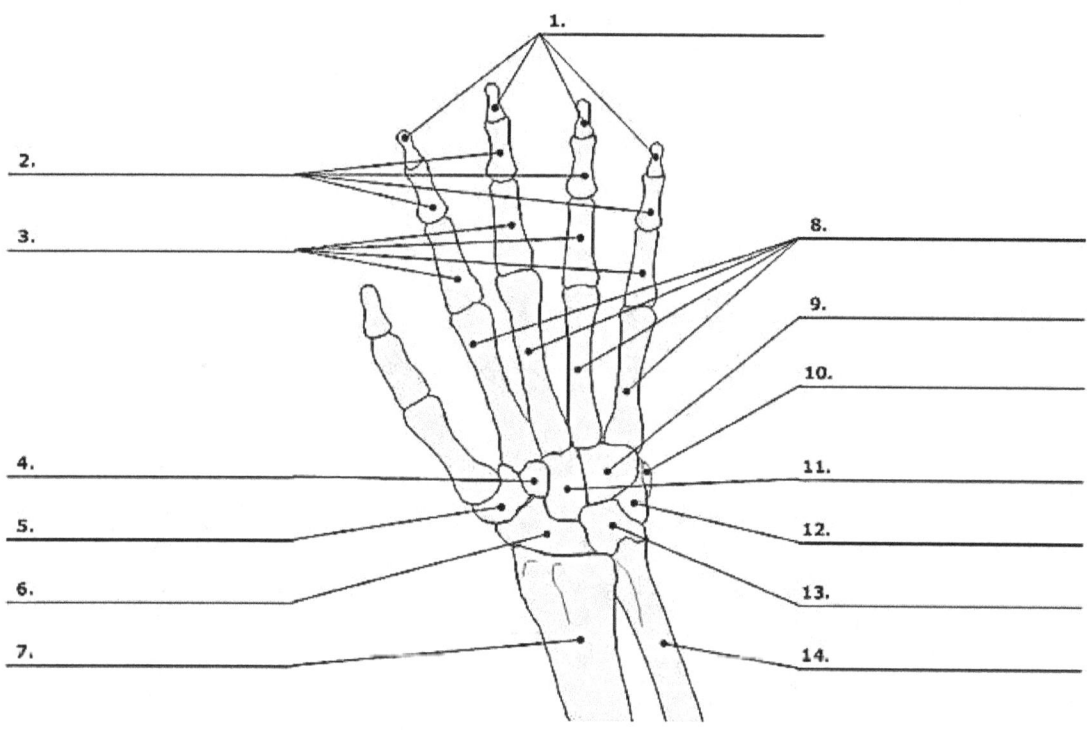

Bones of the Wrist and Hand
Palmar View

1. _____
2. _____
3. _____
4. _____
5. _____
6. _____
7. _____
8. _____
9. _____
10. _____
11. _____
12. _____
13. _____
14. _____

Bones of the Upper Limb

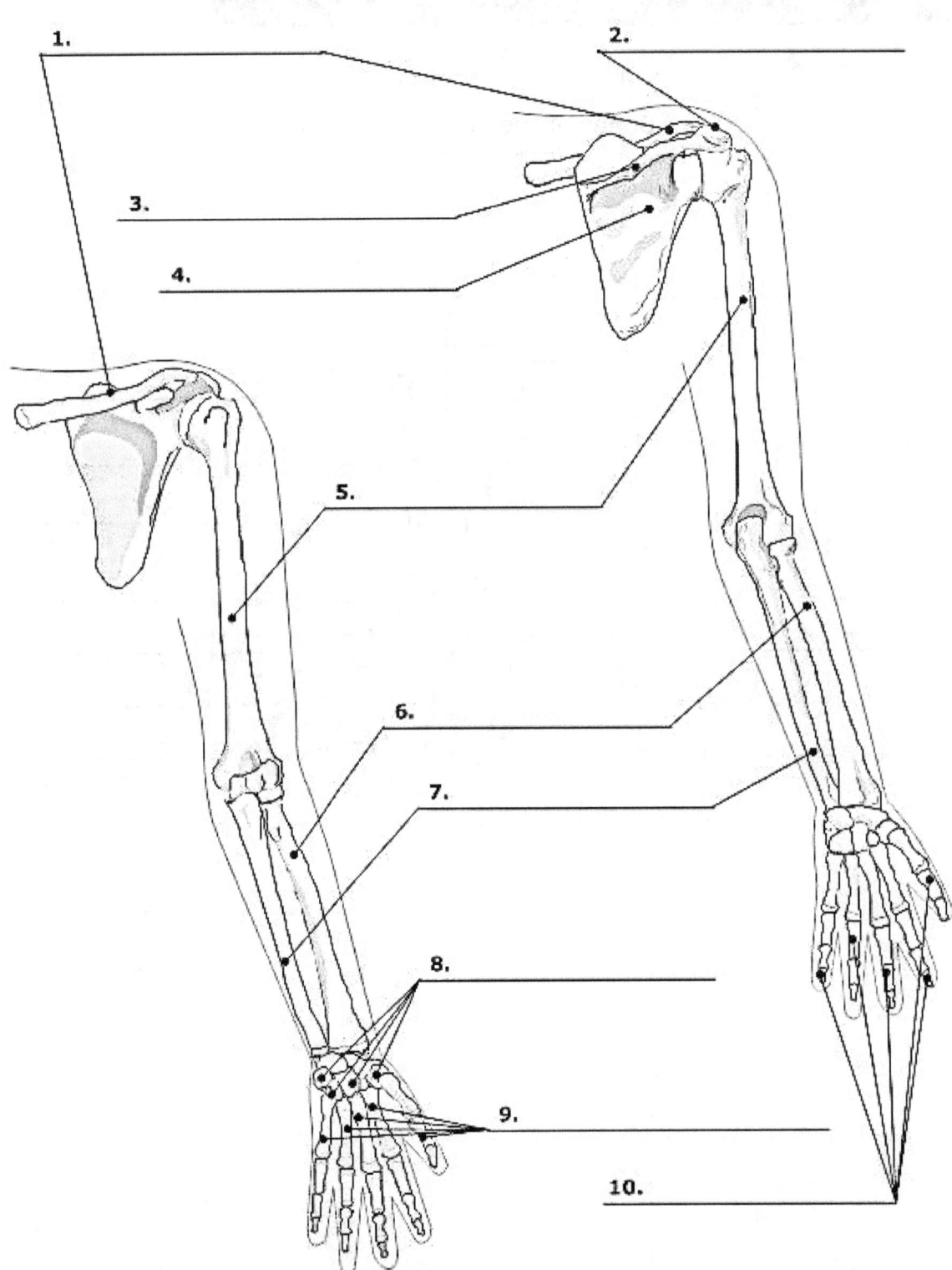

1. _____
2. _____
3. _____
4. _____
5. _____
6. _____
7. _____
8. _____
9. _____
10. _____

116

Bones of the Pelvis

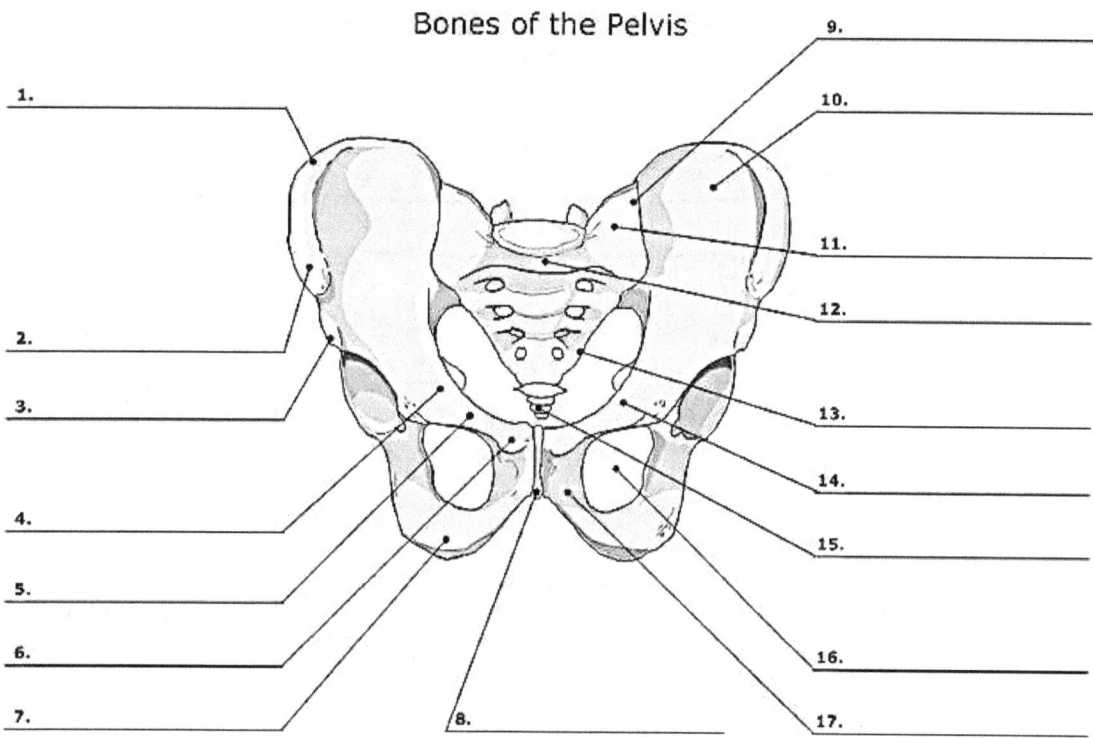

1. _____
2. _____
3. _____
4. _____
5. _____
6. _____
7. _____
8. _____
9. _____
10. _____
11. _____
12. _____
13. _____
14. _____
15. _____
16. _____
17. _____

Bones of the Lower Limb
Femur

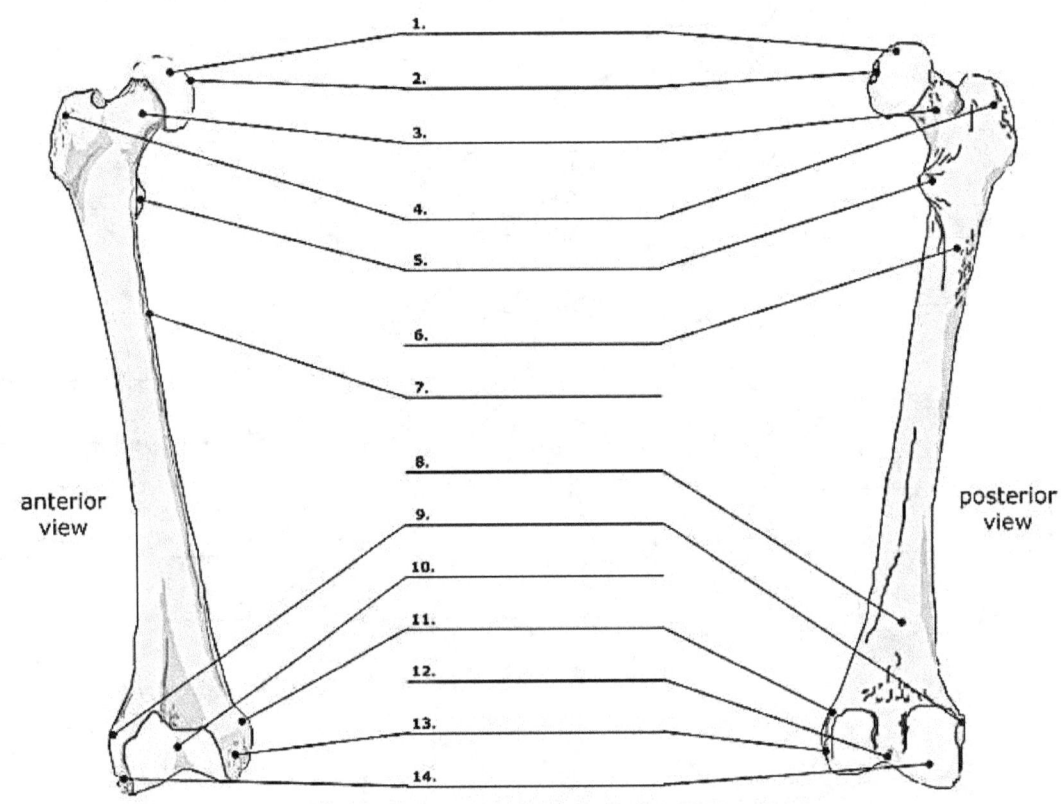

1. _____
2. _____
3. _____
4. _____
5. _____
6. _____
7. _____
8. _____
9. _____
10. _____
11. _____
12. _____
13. _____
14. _____

anterior view

posterior view

Bones of the Lower Limb

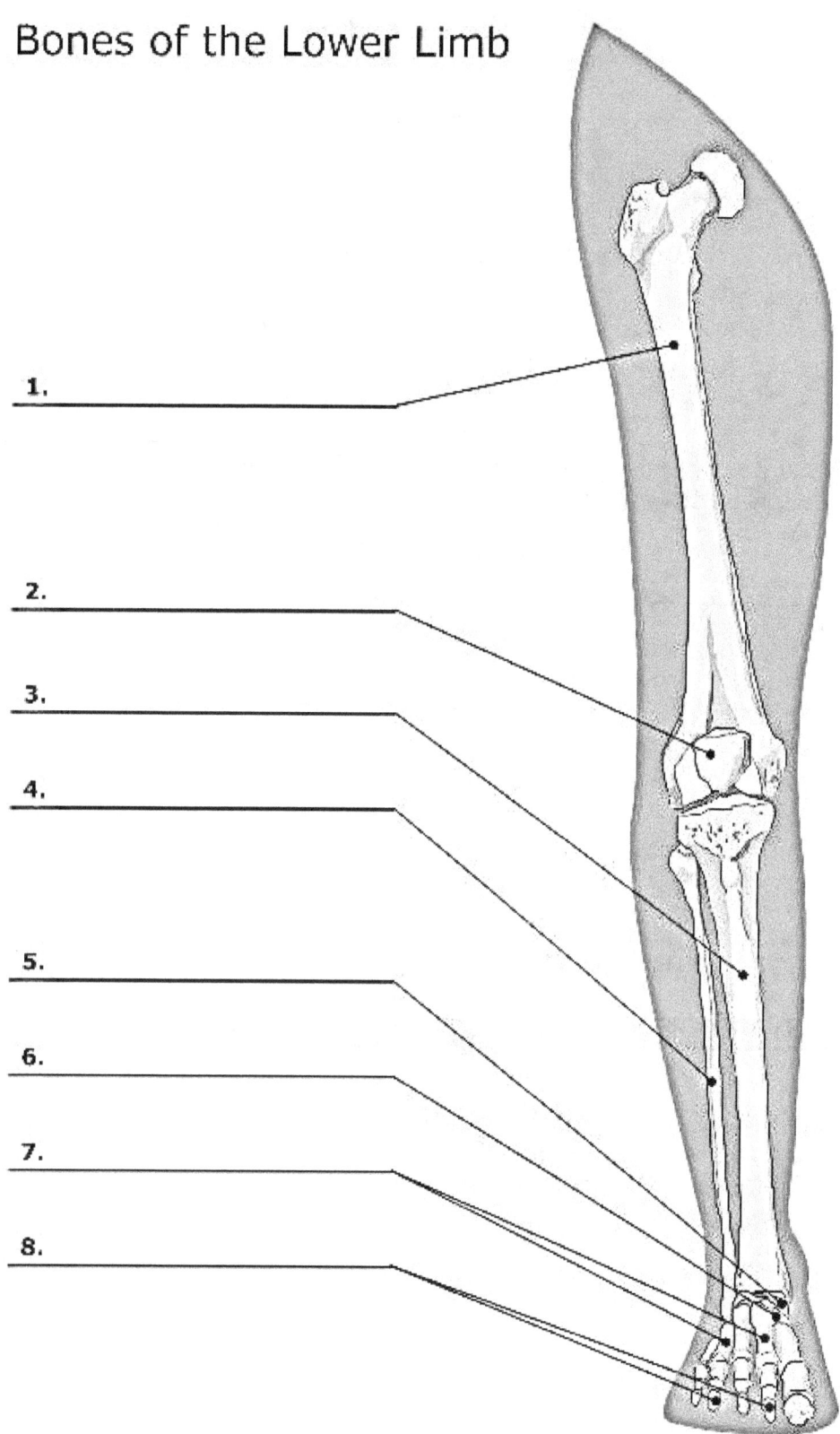

1. _____
2. _____
3. _____
4. _____
5. _____
6. _____
7. _____
8. _____

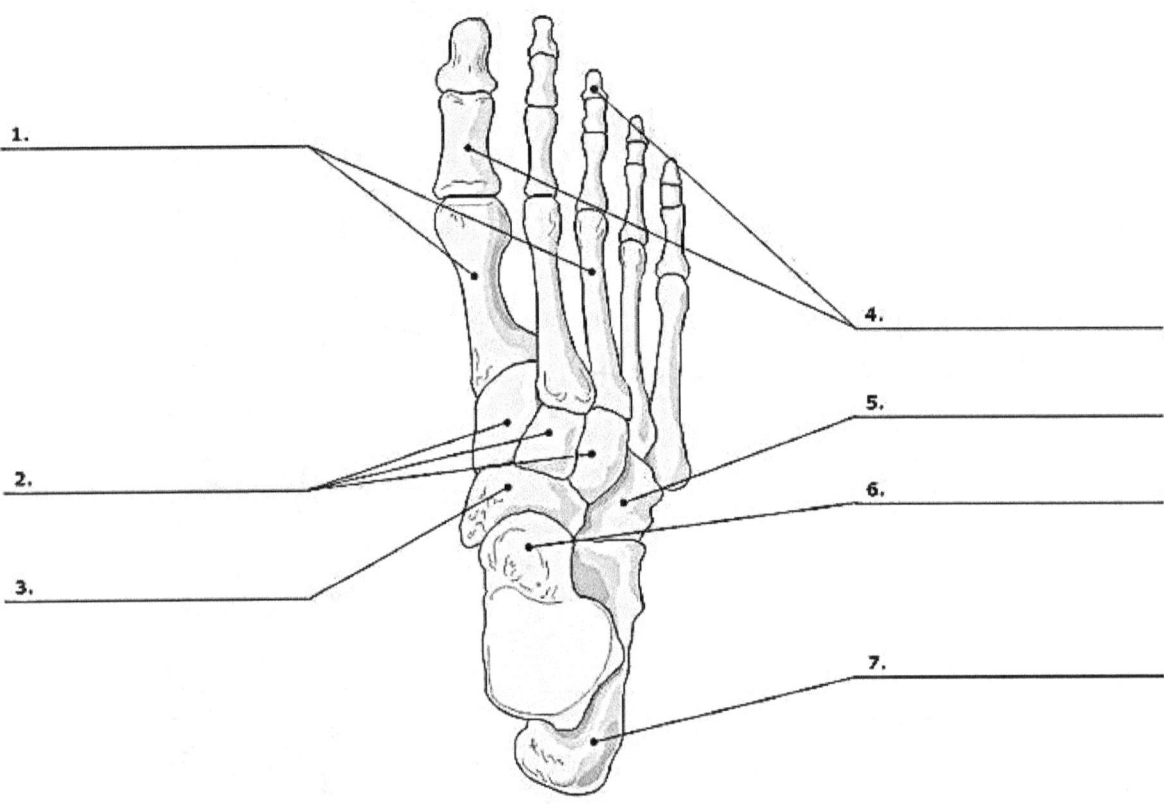

Bones of Ankle and Foot

1. _____
2. _____
3. _____
4. _____
5. _____
6. _____
7. _____

EXERCISE 7.9
EXAMINING THE STRENGTH OF BONE

Purpose of exercise: To observe what happens when bones lose their strength or flexibility.

Bones are composed of a variety of substances. Calcium compounds are found in the hard, outer layer of bones. Calcium provides for the toughness and strength of bones. Bone cells remove calcium phosphate and calcium carbonate from the blood to form bone in a process called ossification.

Materials
- Six chicken bones (*can be from cooked specimen*)
- Six mason jars
- Goggles
- Gloves
- Lab apron
- Masking tape
- Marker
- Forceps or (Tweeters)
- Isopropyl alcohol

- Bleach
- Vinegar
- Coke
- Water
- Paper towels

Procedures

1. Put on goggles, gloves, and lab apron before you begin this experiment. Using the masking tape and marker, label each jar with a different liquid, as shown in the figure. The sixth jar should be labeled "Control."

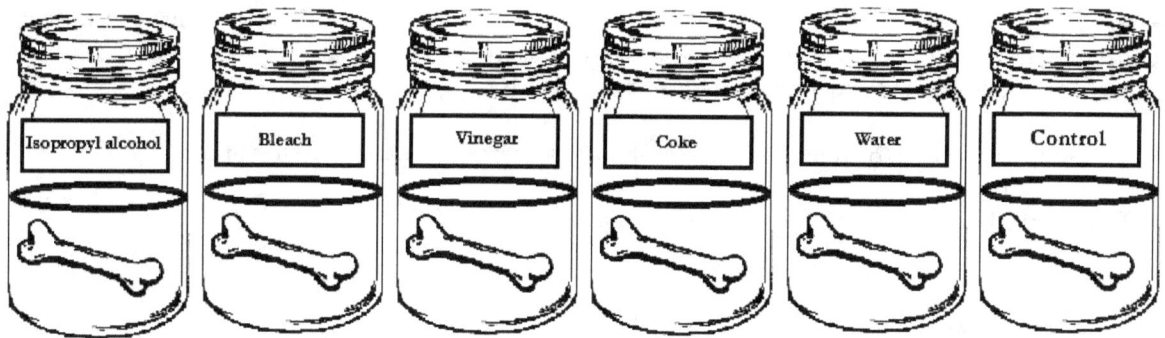

2. Test the strength of each chicken bone by twisting and bending it. Be careful not to snap the bone. Record your results in the data table. Write "no" if the bone does not bend or twist and "yes" if it does.
3. Place one bone in each jar.
4. Fill each jar with the liquid shown on the label. Each bone should be covered by liquid.
5. After five days, remove the bones from the liquids with the forceps. Rinse each bone with water. Then blot dry with paper towels.
6. Retest each bone for strength by twisting and bending. Record your results in the data table.

Data Table

Liquid	Before	After
Isopropyl alcohol		
Bleach		
Vinegar		
Coke		
Water		
Control		

Questions
1. Was there any evidence that calcium was removed from any of the bones? If so, what was it?
2. From which liquids was the calcium removed?
3. What do the liquids that remove calcium have in common?

EXERCISE 7.10
SKELETAL PALPATION

Purpose of exercise: To identify different bones of the skeleton and their markings through touch.

Error! Hyperlink reference not valid.
It is possible to identify most of the bones of the skeleton by palpating, or "feeling" them through the skin. During this exercise, you will be using your fingers to identify different bones of your skeleton. Many of the bones have identifying marks, or more specific parts with specific names, called bony landmarks. Follow the directions for each of the exercises, and using your book, answer the questions about the different bones.

1. Feel for the bump on the back of your head, it feels almost like a ridge just above your hairline. This is the most posterior part of your skull.
    a) Identify the bone of the skull that you are touching.

    b) More specifically, what bony landmark are you touching?

2. Your eye sits in a socket. This socket is protected by a "frame" of bones. Feel the ridge of bone that is directly lateral to your eye. What two facial bones form this ridge?

3. Your jaw easily opens and closes because of a joint that you can feel directly in front of your ear. Place your fingers directly anterior to your ear and open and close your mouth.
    a) What is the anatomical name of the bottom jaw bone?

    b) What is the name of the joint that allows you to open and close your mouth?

4. Directly posterior to your earlobe there is a little pit or depression. Directly behind that is a bony bump.
    a) What is this bony bump called?

    b) Name the bone of the skull on which this bump is located.

5. Your collar bones are the two bones that are horizontal across the front of your shoulders. Using your fingers, trace along this bone to feel where is starts and ends.
    a) What is the anatomical name for this bone?

    b) Your two collar bones connect to another bone in the center of your chest. What is the name of this bone?

6. Feel along the bottom of your ribs on the anterior side of your body. They start low on the sides of your body and come up toward the center of your body.
    a) What number rib is this?

    b) This rib is considered a "false rib". What is a false rib?

7. Feel along your hip bones. They can be felt anteriorly and laterally. What is the anatomical name for these bones?

8. Right underneath your buttocks, you can feel two bony bumps. (These are the ones that hurt when you ride a bike for too long.) What bone are these two bumps a part of?

9. Start at the base of your skull and move your hands slowly down the back of your neck along your spine. The first bump you feel is a cervical vertebra. Which cervical vertebra is this?

10. With one of your hands, reach over your opposite shoulder to feel your shoulder blade. If you bend your elbow and move your arm as if you are running, you can feel this bone moving.
    a) What is the anatomical name of this bone?

    b) What is the bony ridge of this bone called?

11. Where your arm meets your hand at the very base of your arm (distal) on the thumb side, there is a bony bump. What bone are you feeling?

12. Your elbow is a very prominent bony landmark.
    a) What is the anatomical name for your elbow?

    b) What arm bone is the elbow a part of?

13. On one of your hands, feel one of the knuckles that is between your hand and your finger. This is the most proximal knuckle. This joint is formed by two bones. What two bones form this joint?

14. With your leg straight and relaxed, you can move your kneecap around using your thumb and forefinger. What is the anatomical name for your kneecap?

15. Medial to your kneecap (the inside of your leg), you can feel another bony bump.
    a) What is this bony bump called?

    b) What bone is this bony landmark a part of?

16. Move down your leg and feel the bone that forms your ankle on the lateral side.
    a) What is the anatomical name for this very prominent bony landmark?

    b) What bone is this bony landmark located on?

17. Another prominent bony landmark on the lower leg is your shin. Feel along your shin starting below your knee and moving distally.
    a) What is the bone that you are feeling?

    b) More specifically, this is a bony landmark on this bone. What is the name of this bony landmark?

18. Take off your shoes. On the back of your foot is a very bony prominence, your heel, located below the Achille's tendon.
    a) What is the anatomical name for your heel bone?

    b) The heel bone is part of a group of foot bones. What is the name of this group of bones?

EXERCISE 7.11
HANDS-ON WITH THE DISARTICULATED AXIAL SKELETON

Click on *Exercise 7.11* within your online platform or enter the address below into your web browser:
http://www.eskeletons.org/boneviewer/nid/12537/region/skull/bone/cranium

Familiarize yourself with the virtual skeleton at this link. Use the axial skeleton tables as a checklist to learn the different parts. Choose and view different parts of the skeleton by clicking on the skeleton model located on the left of your screen. Place a check next to the bone and boney marking once you've identified the part. The link you've been provided will start you off looking at the human skull. If you get lost, you can return to the www.eskeletons.org home page and look under the **TAXON: HUMAN male, adult**.

Axial Skeleton (Table 1)

I. Cranial Bones (8) ----- (A. – F.)

A. Parietal Bone (2)
1. Temporal line of parietal bone

B. Frontal Bone (1)
1. Supercilliary Arch
2. Supraorbital notch (supraorbital foramen)
3. Supraorbital margin
4. Squamous part of frontal bone
   a. Glabella

C. Ethmoid Bone (1)
1. Crista Galli
2. Cribriform Plate
3. Superior Nasal Concha
4. Middle Nasal Concha
5. Perpendicular plate

D. Occipital Bone (1)
1. Occipital Condyle
2. External Occipital Crest
3. External Occipital Protuberance
4. Superior Nuchal line
5. Inferior Nuchal Line
6. Groove for transverse dural sinus

E. Sphenoid Bone (1)
1. Sella Turcica
   a. Hypophyseal fossa
   b. Dorsum sellae
2. Greater Wings
3. Lesser Wings
4. Pterygoid Processes

F. Temporal Bone (2)
1. External Acoustic Meatus
2. Internal Acoustic Meatus
3. Mandibular Fossa of temporal bone
4. Mastoid Process

II. Foramen/openings in skull
A. Olfactory/Cribriform Foramina
B. Optic foramen & optic Canal
C. Foramen ovale
D. Foramen spinosum
E. Foramen rotundum
F. Foramen Lacerum
G. Carotid Canal
H. Foramen Magnum
I. Hypoglossal canal
J. Jugular Foramen
K. Mandibular Foramen
L. Mental Foramen
M. Stylomastoid Foramen
N. Superior orbital fissure
O. Inferior Orbital Fissure
P. Infraorbital Foramen
Q. Incisive fossa (foramen)
R. Greater palatine foramen

III. Facial Bones (14) ----- (A. – H.)

A. Inferior nasal Concha (2)

B. Nasal Bone (2)

C. Vomer Bone (1)

D. Palatine (2)
1. Horizontal Plate (lamina) of palatine bone
   a. Posterior nasal spine

E. Lacrimal Bone (2)
1. Lacrimal Fossa

F. Maxilla (2)
1. Alveolar Margins of the Maxilla
2. Alveolar Sockets
3. Palatine process of maxilla

    5. Styloid Process
    6. Mastoid incisures
    7. Zygomatic Process of temporal bone
    8. Petrous Part
    9. Groove for sigmoid sinus

G. Mandible (1)
    1. Alveolar Margins of the Mandible
    2. Mandibular symphysis
    3. Mandibular Body
    4. Mandibular Angle
    5. Mandibular Ramus
    6. Mandibular Condyle
    7. Coronoid Process
    8. Mandibular Notch
    9. Mental protuberance
    10. Mental tubercle

H. Zygomatic Bone (2)
    1. Temporal Process of zygomatic bone

Axial Skeleton *(Table 2)*

IV. Orbital surfaces
    A. Orbital surface of the frontal bone
    B. Orbital surface of the sphenoid bone
    C. Orbital surface of the ethmoid bone
    D. Orbital surface of the maxilla bone
    E. Orbital surface of the zygomatic bone

V. Sinuses
    A. Frontal sinus
    B. Ethmoid sinus
    C. Sphenoid sinus
    D. Maxillary sinus

VI. Cranial fossae
    A. Anterior Cranial Fossa
    B. Posterior Cranial Fossa
    C. Middle Cranial Fossa

VII. Sutures
    A. Coronal Suture
    B. Lambdoid Suture
    C. Sagittal Suture
    D. Squamous Suture

VIII. Hyoid Bone
    1. Body
    2. Greater Horn
    3. Lesser Horn

IX. Ribs
    Classification: True Ribs & false ribs
    1. Intercostal Space
    2. Shaft
    3. Costal Angle
    4. Neck of rib
    5. Head of rib
    6. Costal Groove
    7. Tubercle
    8. Costal Cartilage

X. Sternum
    1. Manubrium
        a. Jugular Notch
        b. Clavicular Notch
    2. Body
    3. Xiphoid Process
    4. Sternal Angle
    5. Xiphisternal Joint (Angle)

Axial Skeleton (Table 3)

XII. Vertebrae
A. Intervertebral Foramina
B. Intervertebral Disc

*(General Vertebrae Landmarks)
1. Body (Centrum)
2. Vertebral arch
   a. Pedicle
   b. Lamina
3. Spinous Process
4. Superior articulating process & facet
5. Inferior articulating process & facet
6. Transverse Process
7. Vertebral Foramen/canal
8. Intervertebral foramen (foramina)

Cervical Vertebrae
1. Bifid Spinous Process
2. Vertebral Prominence
3. Transverse Process
4. Transverse Foramina

Atlas
1. Anterior Arch
2. Anterior Tubercle
3. Lateral Masses
   *Superior Articular Facet
   *Inferior Articular Facet
4. Posterior Arch
5. Posterior Tubercle

Axis
1. Body
2. Dens (Odontoid Process)

Thoracic Vertebrae (T1 - T12)
1. Transverse costal facet
2. Superior costal demifacet
3. Inferior costal demifacet

Lumbar Vertebrae (L1 - L5)

Sacrum
A. Sacral Promontory
B. Transverse ridges (Lines)
C. Sacral Foramina
D. Apex
E. Ala of sacrum
F. Auricular Surface
G. Sacral Canal
H. Sacral hiatus
I. Superior Articular Process (of sacrum)
J. Medial Sacral Crest
K. Lateral Sacral Crest

Coccyx

EXERCISE 7.12
HANDS-ON WITH THE DISARTICULATED APPENDICULAR SKELETON

Click on *Exercise 7.12* within your online platform or enter the address below into your web browser:
http://www.eskeletons.org/boneviewer/nid/12537/region/skull/bone/cranium

Familiarize yourself with the virtual skeleton at this link. Use the appendicular skeleton tables as a checklist to learn the different parts. Choose and view different parts of the skeleton by clicking on the skeleton model located on the left of your screen. Place a check next to the bone and boney marking once you've identified the part. The link you've been provided will start you off looking at the human skull. If you get lost, you can return to the www.eskeletons.org home page and look under the **TAXON: HUMAN male, adult**.

Appendicular Skeleton *(Table 1)*	
I. Clavicle 　1. Acromial End 　2. Conoid Tubercle 　3. Shaft 　4. Sternal End	**I. Radius** 　1. Head of Radius 　2. Neck 　3. Radial Tuberosity 　4. Styloid Process of radius 　5. Ulnar Notch of radius
II. Scapula 　1. Borders 　　A. Superior border (superior margin) 　　B. Medial border (vertebral margin) 　　C. Lateral border (axillary margin) 　2. Scapular notch (suprascapular notch) 　3. Acromion Process 　4. Coracoid Process 　5. Infraglenoid tubercle 　6. Angles 　　A. Superior angle 　　B. Inferior Angle 　7. Scapular Spine 　8. Glenoid Fossa (cavity) 　9. Fossa 　　A. Subscapular fossa 　　B. Supraspinous Fossa 　　C. Infraspinous Fossa	**V. Ulna** 　1. Olecranon Process 　2. Coronoid Process of ulna 　3. Trochlear Notch of ulna 　4. Radial Notch of ulna 　5. Head of Ulna 　6. Styloid Process
II. Humerus 　a. Head of Humerus 　b. Neck 　　1. Anatomical Neck 　　2. Surgical Neck	**VI. Carpals (8)** 　1. <u>Proximal row (4)</u>: Scaphoid, Lunate, Triquetrum, Pisiform 　2. <u>Distal row (4)</u>: Trapezium, Trapezoid, Capitate, Hamate

c. Tubercle
          1. Greater Tubercle
          2. Lesser Tubercle
    d. Intertubercular groove
    e. Deltoid Tuberosity
    f. Epicondyle
         3. Lateral Epicondyle
         4. .Medial Epicondyle
    g. Fossa
         5. Olecranon Fossa
         6. Radial Fossa
         7. Coronoid Fossa	

Appendicular Skeleton *(Table 2)*	

I. Metacarpals: (Numbered 1-5)
   1. Base
   2. Shaft
   3. Head
	VII. Pubic Bone
	   1. Pubic Body
	   2. Pubic Crest
	   3. Pubic tubercle
	   4. Ramus
	      A. Superior Pubic Ramus
	      B. Inferior Pubic Ramus

II. Phalanges (14)
   1. Proximal Phalanx
   2. Middle Phalanx
   3. Distal Phalanx
	VIII. Femur
	   1. Head of Femur
	      A. Fovea Capitis
	   2. Neck of Femur
	   3. Trochanter
	      A. Greater Trochanter
	      B. Lesser Trochanter
	   4. Trochanteric fossa
	   5. Intertrochanteric Line
	   6. Intertrochanteric Crest
	   7. Gluteal Tuberosity
	   8. Shaft of femur
	   9. Linea Aspera
	   10. Intercondylar fossa
	   11. Condyle
	      A. Lateral Condyle
	      B. .Medial Condyle
	   12. Epicondyle
	      A. Lateral Epicondyle
	      B. Medial Epicondyle
	   13. Patellar surface

III. Coxal bone
   1. Acetabulum
   2. Obturator Foramen
   3. Ilium
      A. Ala
      B. Iliac Crest
      C. Spines
         a. Anterior Superior Iliac Spine
         b. Posterior Superior iliac Spine
         c. Anterior Inferior Iliac Spine
         d. Posterior Inferior Iliac Spine
      E. Auricular Surface (sacroiliac Joint)
      F. Iliac Fossa
      G. Pelvic Brim (Arcuate Line)
      H. Greater Sciatic Notch
	IX. Patella
	   1. Base
	   2. Apex

IV. Ischium
   1. Ischial body
   2. Ischial Spine
   3. Ischial Ramus
   4. Ischial Tuberosity
   5. Lesser Sciatic Notch
	X. Tibia
	   1. Condyle
	      A. Lateral Condyle
	      B. Medial Condyle
	   2. Intercondylar Eminence
	   3. Tibial Tuberosity
	   4. Anterior Crest (shin)
	   5. Shaft of Tibia
	   6. Medial Malleolus
	   7. Articulating Surface for Fibular Head

V. Fibula
   1. Head of Fibula
   2. Shaft of Fibula
   3. Lateral Malleolus
	XI. Tarsals (7)
	   1. Calcaneus
	   2. Talus
	   3. Navicular
	   4. Medial, intermediate, & lateral Cuneiform (3)
	   5. Cuboid

VI. Metatarsals: (Numbered 1-5)
   1. Base
   2. Shaft
   3. Head
	XII. Phalanges (14)
	   1. Proximal Phalanx
	   2. Middle Phalanx
	   3. Distal Phalanx

EXERCISE 7.13
MOVEMENT OF SYNOVIAL JOINTS (PASSIVE VS. ACTIVE RANGE OF MOTION)

Purpose of exercise: To evaluate synovial joint mobility through passive and active range of motion.

Range of motion is the distance or amount of freedom your joint can be moved in a certain direction. Your passive and active range of motion (ROM) may be very different, not only from each other, but also at the joints themselves. Passive range of motion involves someone else moving a joint for you. An example of passive range of motion would be if a healthcare professional were testing your shoulder joint, and moves it without your assistance. Active range of motion means that you are able to move the joint through its range of motion on your own (without manual assistance).

Materials
This exercise will require you to have a lab partner.

Procedure
1. Starting with passive range of motion, you will assess your lab partner as they perform the movements in the Figure below.
2. In the Data Table, record **"positive"** if your partner CAN PERFORM the range of movement and **"negative"** if they CANNOT PERFORM the range of motion.
3. After you have assessed your lab partner's passive range of motion, proceed next to examine their active range of motion.
4. Also, make sure to note the presence of pain or any crackling, grating feeling or sounds you detect in the comments section.

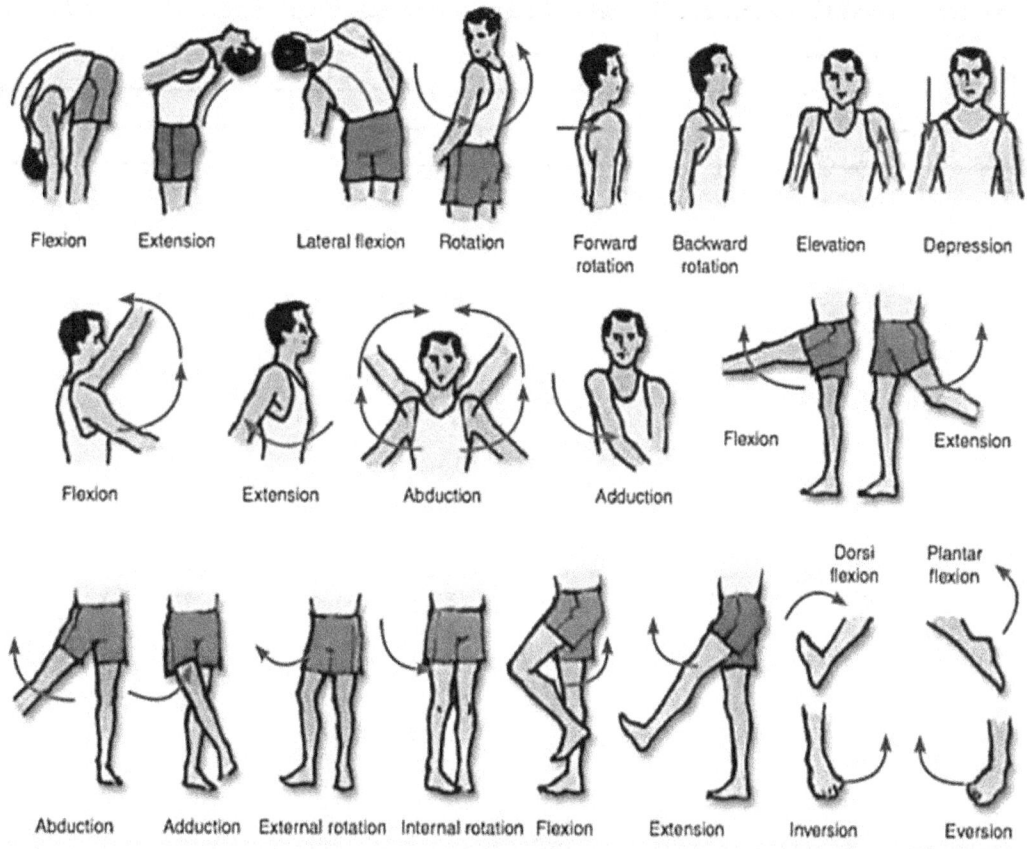

Data Table

Movement	Passive ROM (+/-)	Active ROM (+/-)	Comments
1. Flexion			
2. Extension			
3. Lateral Flexion			
4. Rotation			
5. Forward rotation			
6. Backward rotation			
7. Elevation			
8. Depression			
9. Flexion			
10. Extension			
11. Abduction			
12. Adduction			
13. Flexion			
14. Extension			
15. Abduction			
16. Adduction			
17. External rotation			
18. Internal rotation			
19. Flexion			
20. Extension			
21. Dorsi Flexion			
22. Plantar Flexion			
23. Inversion			
24. Eversion			

EXERCISE 7.14
JOINT QUESTIONS

1. The function of the articular cartilage is
    a. to reduce friction
    b. to prevent the surfaces of the bones in the joint from making contact
    c. to produce a lubricating fluid
    d. to allow growth in length of the bone
    e. a and b are correct

2. Moving the femur forward during walking is an example of
    a. abduction
    b. circumduction
    c. flexion
    d. elevation
    e. adduction

3. When a gymnast performs the "splits" moving the lower limbs laterally, the primary movement at the hip joint is
    a. rotation
    b. adduction
    c. extension
    d. protraction
    e. abduction

4. In the anatomical position, the palms of the hands are
    a. supinated
    b. flexed
    c. pronated
    d. depressed
    e. retracted

5. Nodding your head "yes" in response to a question involves
    a. abduction and adduction
    b. circumduction
    c. protraction and retraction
    d. rotation
    e. flexion and extension

6. Along which side of the shoulder joint will you find the glenohumeral ligaments?
    a. inferior
    b. superior
    c. anterior
    d. lateral

7. The shoulder joint is primarily stabilized by _____.
    a. ligaments and muscles that move the humerus
    b. the scapula
    c. glenohumeral ligaments only
    d. the clavicle
    e. b and c

8. The carpal and metacarpal of the thumb form this type of joint:
    a. hinge
    b. gliding
    c. saddle
    d. ball and socket

9. In a synovial joint, the synovial membrane:
    a. lines the joint capsule
    b. covers the articular cartilage
    c. both A and B
    d. both A and B, and prevents friction

10. Joints
    a. bind bones.
    b. allow bones to grow.
    c. enable body parts to move.
    d. all of the above.

11. Types of fibrous joints include:
    a. synarthrotic, amphiarthrotic, and diarthrotic joints.
    b. syndesmosis, suture, and gomphosis joints.
    c. synchondrosis, symphysis, and synovial joints.
    d. pivot, condylar, and ellipsoidal joints

12. Which of the following is not a type of fibrous joint?
    a. syndesmosis
    b. coronal suture
    c. gomphosis
    d. symphysis

13. A joint capsule is reinforced by:
    a. tendons binding articular ends of bones together.
    b. articular cartilage cushioning ends of bones.
    c. ligaments binding articular ends of bones together.
    d. hyaline cartilage providing strength to the capsule walls.

14. The largest and most complex synovial joint is the
    a. hip joint.
    b. knee joint.
    c. elbow joint.
    d. shoulder joint.

15. The four muscles of the shoulder joint are referred to as the _____.

16. What functional joint class contains the least mobile joints?

17. What does the term arthritis mean?

18. The fluid-filled sac that eases movement of a tendon over a bone is a/an
    _____.

19. Name the three structural classifications of joints.

20. Name the three functional classifications of joints.

EXERCISE 7.15
JOINT ASSESSMENT

Purpose of exercise: To assess the shoulder, knee, and ankle joint functionality using orthopedic tests.

Orthopedic examinations are designed to evaluate individuals for musculoskeletal impairment. Orthopedic tests enable the healthcare professional, to identify a specific area of injury and aid in the diagnosis and treatment plan of the injured patient. There is a general plan for physical assessment that includes taking a patient's history; examining how the patient moves and how individual joints move; evaluating sensation and reflexes; and, if necessary, administering diagnostic tests to aid in the diagnosis.

In this exercise, you and your lab partner will assess the shoulder, knee, and ankle joint functionality using several orthopedic tests. There are numerous orthopedic tests, yet we will only use fifteen in this exercise. One person will act as the Healthcare provider and the other will serve as the patient. Please place a positive or negative symbol below the **Test Results** heading next to the test once you have completed assessing the patient using that test. A positive symbol indicates one of the following: instability, decreased range of motion, hypermobility, injury, weakness, further evaluation, or apprehension/pain at the affected area. A negative symbol indicates that the area in question is within normal limits.

Materials:
- This exercise will require you to have a lab partner.
- Therapy table or bed *(You also have the option to position your patient on the floor, however please make sure that the area is clean and the patient is comfortable)*.
- Chair

Data Table

Area of Assessment	Name of Test	How To Perform Test	Positive Sign	Indication	Test Results (-/+)
(5) Shoulder Tests					
Shoulder	**Acromioclavicular Shear**	1. Patient is seated. 2. Healthcare provider stands behind the patient. 3. Healthcare provider will then place cupped hands over the patient's shoulder, the fingers interlaced. (*One palm on the clavicle, the other hand on the scapula*) 4. Healthcare provider will now slowly squeeze the heels of his or her hands together.	Pain or excessive movement of the acromioclavicular joint	Indicate an acromioclavicular joint ligament sprain	
Shoulder	**Crank Test**	1. Patient seated with shoulder abducted 180° & elbow flexed 90° 2. Healthcare provider applies long axis compression & rotates humerus internally & externally.	Patient expresses pain or increased motion	Tests the integrity of the glenoid labrum; also for labral tear. May also produce an audible or palpable clicking in the glenohumeral joint.	

Shoulder	**Posterior Apprehension**	1. Patient is seated. 2. Healthcare provider elevates the patient's shoulder in the plane of the scapula to 90° while using the other hand to stabilize the scapula. 3. Healthcare provider then applies a force posterior on the patients elbow while horizontally adducting and internally rotating the arm.	Patient expresses apprehension and/ or might try to move their affected shoulder away from the pressure. Or increase motion.	Tests for dislocation or posterior instability of the humerus	
Shoulder	**Jobes Relocation Test**	1. Patient is supine. 2. Healthcare provider pre-positions the shoulder at 90° of abduction and maximal external rotation. 3. Healthcare provider will now grasps the patient's wrist and hand with his/her distal hand while applying a posterior force to the humeral head while externally rotating the shoulder	Administered after the Apprehension test produces a positive result. if the patient's apprehension or pain is reduced in this position, the Jobes Relocation test is considered to be positive.	Tests for possible glenohumeral instability, dislocation and subluxation	

Shoulder	**Anterior Drawer**	1. Patient is supine. 2. Healthcare provider stabilizes the anterior distal leg with one hand & grasps the patient's calcaneus and rear foot with their second hand. 3. Healthcare provider then places the patient's foot into 10-15 degrees of plantar flexion and translates the rear foot anteriorly.	the talus translates forward	Test ligamentous laxity or instability of the Anterior Talofibular Ligament.	
		(7) Knee Tests			
Knee	**Valgus Stress**	1. Patient is supine 2. Healthcare provider will hold patient's leg with knee extended and relaxed. 3. Healthcare provider will now apply lateral to medial pressure to the patient's lateral knee.	Pain or hyper mobility at medial aspect of knee.	MCL Sprain or Instability	
Knee	**Varus Stress**	1. Patient is supine 2. Healthcare provider will hold patient's leg with knee extended and relaxed. 3. Healthcare provider will now apply medial to lateral pressure to the patient's medial knee.	Pain or hyper mobility at lateral aspect of knee.	LCL Sprain or Instability	

Knee	**Lachman**	1. Patient is supine 2. Healthcare provider will hold patient's leg with patient's knee flexed between 10 to 30 degrees. 3. Healthcare provider will now pull the proximal tibia of the patient forward.	Pain or excessive anterior motion of the tibia, and disappearance of the infrapatellar tendon slope.	ACL Sprain	
Knee	**Hyperflexion Compression**	1. Patient is prone. 2. Patient will then flex affected knee to 130°-150° 3. Healthcare provider's one hand grasps patient's heel and ankle while the other hand stabilizes the leg. 4. Healthcare provider will applies downward force by pushing the patient's foot and tibia down into the table while rotating the foot internally and externally.	Pain with the movement	Meniscus Injury or Lesion	

Knee	**Apley's Distraction**	1. Patient is prone, with their affected knee flexed 90° 2. Healthcare provider places their own knee on patient's posterior thigh to stabilize 3. Healthcare provider grasps patient's leg proximal to the ankle 4. Healthcare provider applies traction to the tibia towards the ceiling (this distracts the knee joint) –then apply internal and external rotation of the tibia while tractioning.	1. Pain on the medial side. 2. Pain on the lateral side.	1. Medial collateral ligament damage/ injury 2. Lateral collateral ligament damage/ injury	
Knee	**Patellar Apprehension Test**	1. Patient is supine with their affected knee extended 2. Healthcare provider uses a slow and moderate pressure against the medial aspect of the patella moving it in a lateral direction 3. Healthcare provider observes patient's reaction	Patient expresses apprehension and/ or might try to move their affected knee away from the pressure.	Lateral dislocation of the patella	

Knee	**Posterior Drawer Test**	1. Patient supine with knee bent 90°. 2. Healthcare provider places thumbs over the anterolateral and anteromedial knee joint line. 3. Healthcare provider pushes the tibia posteriorly.	Patient expresses pain or increased motion	PCL sprain or rupture	
(3) Ankle Tests					
Ankle	**Ankle Drawer Test**	1. Patient positions foot in slight plantar flexion 2. Brace anterior shin with left hand 3. Pull heel anteriorly with right hand	1. Anterior motion of the foot: Anterior talofibular ligament instability. 2. Posterior motion of the foot: Posterior ankle ligament instability.	Tests the strength of the Anterior Talofibular Ligament and for the instability of the ankle.	
Ankle	**External Rotational Stress test/ Kleiger Test**	Patient's foot is in neutral position with the lower leg stabilized. Healthcare provider will then externally rotate patient's foot.	Pain will be at site of the anterior tibiofibular ligament.	Tests for sprain, instability or ligament laxity.	

Ankle	Talar Tilt	1. Patient is seated with foot and ankle unsupported. 2. Patient's foot is positioned in 10-20 degrees of plantarflexion. 3. Patient's distal lower leg is stabilized with one hand of the Healthcare provider just proximal to the malleloi and the back of foot is inverted with his/her other hand. 4. The lateral aspect of the talus is palpated to determine if tilting occurs. The laxity is compared to the contralateral side.	Laxity and/or pain	Tests for injury of the Anterior Talofibular ligament and the Calcaneofibular ligament	

EXERCISE 7.16
RADIOLOGICAL INTERPRETATION

Purpose of exercise: To provide students a foundational knowledge in the art of radiological interpretation while reinforcing anatomy and physiology concepts.

During a radiographic procedure, an x-ray beam is passed through the body. A portion of the x-rays are absorbed or scattered by the internal structure and the remaining x-ray pattern is transmitted to a detector so that an image may be recorded for later evaluation. The recoding of the pattern may occur on film or through electronic means. X-rays are a type of radiation called electromagnetic waves. X-ray images show the parts of your body in different shades of black (radiolucent) and white (radiopaque). This is because different tissues absorb different amounts of radiation. Calcium in bones absorbs x-rays the most, so bones look white. Fat and other soft tissues absorb less, and look gray. Air absorbs the least, so lungs look black.

The most familiar use of x-rays is checking for broken bones, but x-rays are also used in other ways. For example, chest x-rays can spot pneumonia. Mammograms use x-rays to look for breast cancer. It is used to diagnose or treat patients by recording images of the internal structure of the body to assess the presence or absence of disease, foreign objects, and structural damage or anomaly.

This website allows students to build their skills in medical x-ray imaging by giving free access to materials. **YOU ARE NOT REQUIRED TO BUY ANYTHING ON THE RADIOLOGYMASTER CLASS WEBSITE FOR THIS COURSE**.

Click on *Exercise 7.16(a)* within your online platform or enter the address below into your web browser before starting this exercise: https://www.radiologymasterclass.co.uk/tutorials/tutorials_howto

Now click on *Exercise 7.16(b)* within your online platform or enter the address below into your web browser: https://www.radiologymasterclass.co.uk/tutorials/musculoskeletal/principles/bones_joints_x-ray_page1

Musculoskeletal X-ray

This tutorial discusses musculoskeletal X-ray anatomy in general terms, and introduces some important concepts regarding musculoskeletal X-ray interpretation. Knowledge of normal bone, joint and soft tissue appearances enables accurate description of abnormalities seen on X-ray. Both normal and abnormal X-rays are used to illustrate key viewing principles. As for all X-rays, a systematic approach is required. By the end of the tutorial you will be familiar with all the important structures of the musculoskeletal system, which should be checked whenever you look at a musculoskeletal X-ray. Before you start, look at the normal musculoskeletal X-ray. Please read and follow the instructions provided on the website. Click the "**Next >>**" button to move through the images on the page.

EXERCISE 7.17
RADIOLOGICAL INTERPRETATION - TRAUMA X-RAY - UPPER LIMB & LOWER LIMB

Purpose of exercise: To provide students a foundational knowledge in the art of radiological interpretation while reinforcing anatomy and physiology concepts.

During a radiographic procedure, an x-ray beam is passed through the body. A portion of the x-rays are absorbed or scattered by the internal structure and the remaining x-ray pattern is transmitted to a detector so that an image may be recorded for later evaluation. The recoding of the pattern may occur on film or through electronic means. X-rays are a type of radiation called electromagnetic waves. X-ray images show the parts of your body in different shades of black (radiolucent) and white (radiopaque). This is because different tissues absorb different amounts of radiation. Calcium in bones absorbs x-rays the most, so bones look white. Fat and other soft tissues absorb less, and look gray. Air absorbs the least, so lungs look black.

The most familiar use of x-rays is checking for broken bones, but x-rays are also used in other ways. For example, chest x-rays can spot pneumonia. Mammograms use x-rays to look for breast cancer. It is used to diagnose or treat patients by recording images of the internal structure of the body to assess the presence or absence of disease, foreign objects, and structural damage or anomaly.

This website allows students to build their skills in medical x-ray imaging by giving free access to materials. **YOU ARE NOT REQUIRED TO BUY ANYTHING ON THE RADIOLOGYMASTER CLASS WEBSITE FOR THIS COURSE**.

Click on *Exercise 7.17(a)* within your online platform or enter the address below into your web browser before starting this exercise: https://www.radiologymasterclass.co.uk/tutorials/tutorials_howto

Now click on *Exercise 7.17(b)* within your online platform or enter the address below into your web browser: https://www.radiologymasterclass.co.uk/tutorials/musculoskeletal/trauma/trauma_x-ray_start

Now click on *Exercise 7.17(c)* within your online platform or enter the address below into your web browser: https://www.radiologymasterclass.co.uk/tutorials/musculoskeletal/x-ray_trauma_upper_limb/upper_limb_trauma_x-ray_start

Now click on *Exercise 7.17(d)* within your online platform or enter the address below into your web browser: https://www.radiologymasterclass.co.uk/tutorials/musculoskeletal/x-ray_trauma_lower_limb/lower_limb_trauma_x-ray_start

Trauma X-ray
This tutorial introduces some key principles regarding the role of X-rays in the setting of trauma. The common terminology used for describing fractures is discussed. Basic concepts are introduced regarding the use of X-rays in management of fractures and dislocations. X-ray appearances of fracture complications, and some common fracture mimics are also described. Please read and follow the instructions provided on the website. Click the "**Next >>**" button to move through the images on the page.

EXERCISE 7.18
RADIOLOGICAL INTERPRETATION – AXIAL SKELETON

Purpose of exercise: To provide students a foundational knowledge in the art of radiological interpretation while reinforcing anatomy and physiology concepts.

During a radiographic procedure, an x-ray beam is passed through the body. A portion of the x-rays are absorbed or scattered by the internal structure and the remaining x-ray pattern is transmitted to a detector so that an image may be recorded for later evaluation. The recoding of the pattern may occur on film or through electronic means. X-rays are a type of radiation called electromagnetic waves. X-ray images show the parts of your body in different shades of black (radiolucent) and white (radiopaque). This is because different tissues absorb different amounts of radiation. Calcium in bones absorbs x-rays the most, so bones look white. Fat and other soft tissues absorb less, and look gray. Air absorbs the least, so lungs look black.

The most familiar use of x-rays is checking for broken bones, but x-rays are also used in other ways. For example, chest x-rays can spot pneumonia. Mammograms use x-rays to look for breast cancer. It is used to diagnose or treat patients by recording images of the internal structure of the body to assess the presence or absence of disease, foreign objects, and structural damage or anomaly.

This website allows students to build their skills in medical x-ray imaging by giving free access to materials. <u>**YOU ARE NOT REQUIRED TO BUY ANYTHING ON THE RADIOLOGYMASTER CLASS WEBSITE FOR THIS COURSE**</u>.

Click on *Exercise 7.18(a)* within your online platform or enter the address below into your web browser before starting this exercise: **https://www.radiologymasterclass.co.uk/tutorials/tutorials_howto**

Now click on *Exercise 7.18(b)* within your online platform or enter the address below into your web browser: https://www.radiologymasterclass.co.uk/tutorials/musculoskeletal/trauma/trauma_x-ray_start

Now click on *Exercise 7.18(c)* within your online platform or enter the address below into your web browser: https://www.radiologymasterclass.co.uk/tutorials/musculoskeletal/x-ray_trauma_spinal/x-ray_fracture_start

Trauma X-ray

This tutorial discusses the X-ray appearances of trauma to the axial skeleton. Examples of common injuries seen on X-ray are shown, with normal images for comparison. **Please read and follow the instructions provided on the website.** Click the "**Next >>**" button to move through the images on the page.

MUSCULAR SYSTEM
LAB 8

CRASHCOURSE VIDEO(S)

Click on the video embedded within your online platform or enter the address below into your web browser:
1. **https://youtu.be/Ktv-CaOt6UQ**
2. **https://youtu.be/I80Xx7pA9hQ**

(Please watch the videos below before continuing)

DEFINING KEY TERMS:

1. A-Band:

2. Acetylcholine (ACh):

3. Acetylcholinesterase:

4. Actin:

5. Aerobic Respiration:

6. Anaerobic Respiration:

7. Aponeurosis:

8. Calmodulin:

9. Citric Acid Cycle

10. Contraction:

11. Depolarization:

12. Electron Transport Chain:

13. Endomysium:

14. Epimysium:

15. Fascia:

16. Glycolysis:

17. I-Band:

18. Insertion:

19. Isometric Contraction:

20. Isotonic Contractions:

21. Lactic Acid:

22. Motor Neuron:

23. Muscle Fatigue:

24. Muscle Tone:

25. Myosin:

26. Neuromuscular Junction:

27. Neurotransmitter:

28. Origin:

29. Perimysium:

30. Peristalsis:

31. Polarization:

32. Prime Mover (agonist):

33. Sarcomeres:

34. Sarcoplasm:

35. Sarcoplasmic Reticulum (SR):

36. Striations:

37. Synergist:

38. Tendon:

39. Thick Filaments:

40. Thin Filaments:

41. Transverse Tubules (TT):

42. Tropomyosin:

43. Troponin:

44. Z-line:

EXERCISE 8.1
INTERACTIVE HUMAN SKELETAL MUSCLE DIAGRAM

Purpose of exercise: To identify the names and location of human skeletal muscles.

Click on *Exercise 8.1(a)* within your online platform or enter the address below into your web browser:
https://www.getbodysmart.com/ap/muscularsystem/menu/menu.html

Familiarize yourself with the interactive human skeletal muscle diagram at this link. Choose and view different parts of the muscular system by clicking on the **Muscles Tutorials** (organized by action sites) link located on the right of your screen.

Now click on *Exercise 8.1(b)* within your online platform or enter the address below into your web browser:
https://www.biodigital.com/

- *(This is a free site, but you will need to sign up with your name and a validated email address. You can use your personal email address or school email address. MAKE SURE TO CREATE A PASSWORD YOU CAN REMEMBER. I WOULD SUGGEST WRITING IT DOWN AND KEEPING IT IN A SAFE PLACE. YOU WILL USE IT AGAIN.)*

Click **SIGN UP**. Once you have provided the appropriate information, you are now able to access BioDigital's website content. Click **LOG IN** using the email and password you created. Then click **SIGN IN**. On the left of your screen choose from the systems listed.

Please click on the square boxes next to this icon. [Search Human Library]. Now click on the **Anatomy By Systems** icon. [icon] Click on the following image(s) with caption. **Male Muscular System, Female Muscular System.** View the structures associated with this system.

EXERCISE 8.2
IDENTFYING STRUCTURE OF SKELETAL MUSLCES (AND ASSOCIATED CONNECTIVE TISSUE)

Purpose of exercise: To identify structural components of skeletal muscle tissues.

Name the parts of the skeletal muscle on the next page

Muscle
A.
B.
C.
D.
E.

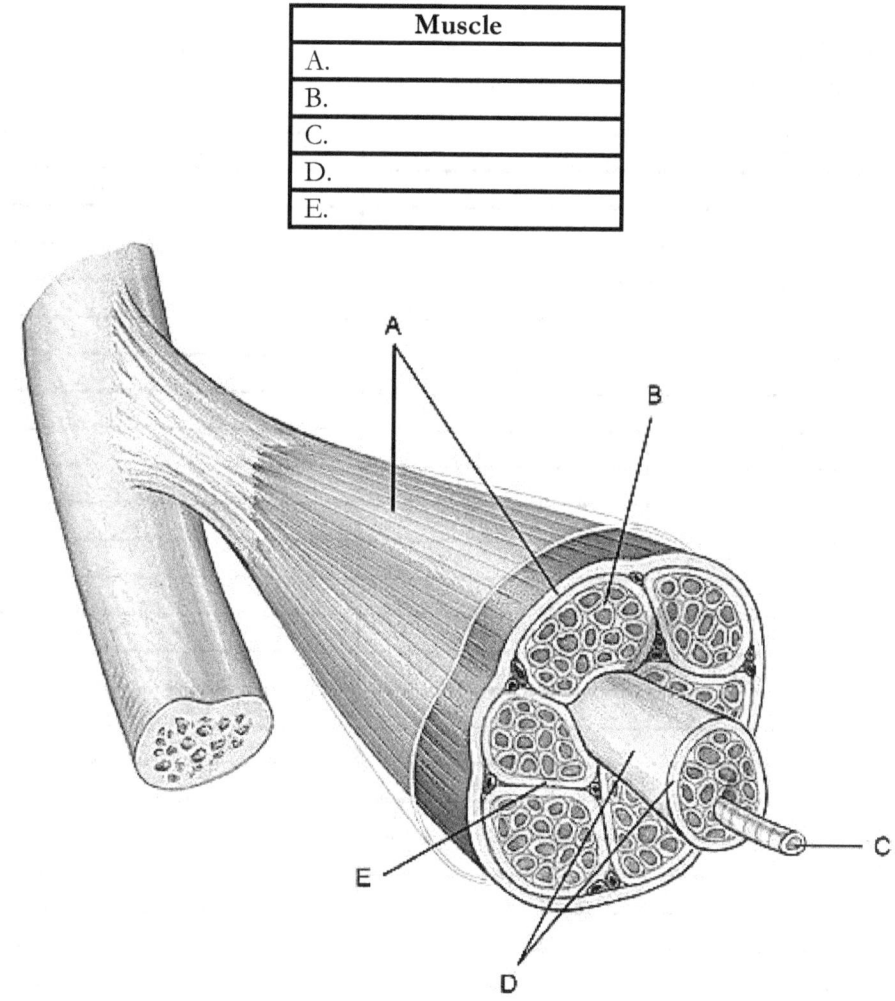

EXERCISE 8.3
IDENTFYING SKELETAL MUSLCES

Purpose of exercise: To identify the names, origin, insertion, and action of human skeletal muscles.

Click on *Exercise 8.3* within your online platform or enter the address below into your web browser:
http://www.med.umich.edu/lrc/coursepages/m1/anatomy2010/html/anatomytables/muscles_alpha.html

Review the list of skeletal muscles provided on the page. Scroll through the list of muscles t6o further your understanding.

EXERCISE 8.4
COLORING SKELETAL MUSLCES

Purpose of exercise: To identify the names, origins, and insertions of superficial human skeletal muscles.

In this exercise you will color the superficial muscles, name each muscle, and record the origin and insertion of the entire body. Please remember to write the color used to identify each muscle.

Muscle	Origin	Insertion	Color
1.			
2.			
3.			
4.			
5.			
6.			
7.			
8.			
9.			
10.			
11.			
12.			
13.			
14.			
15.			
16.			
17.			
18.			
19.			
20.			
21.			
22.			
23.			
24.			
25.			
26.			
27.			
28.			
29.			
30.			
31.			
32.			
33.			

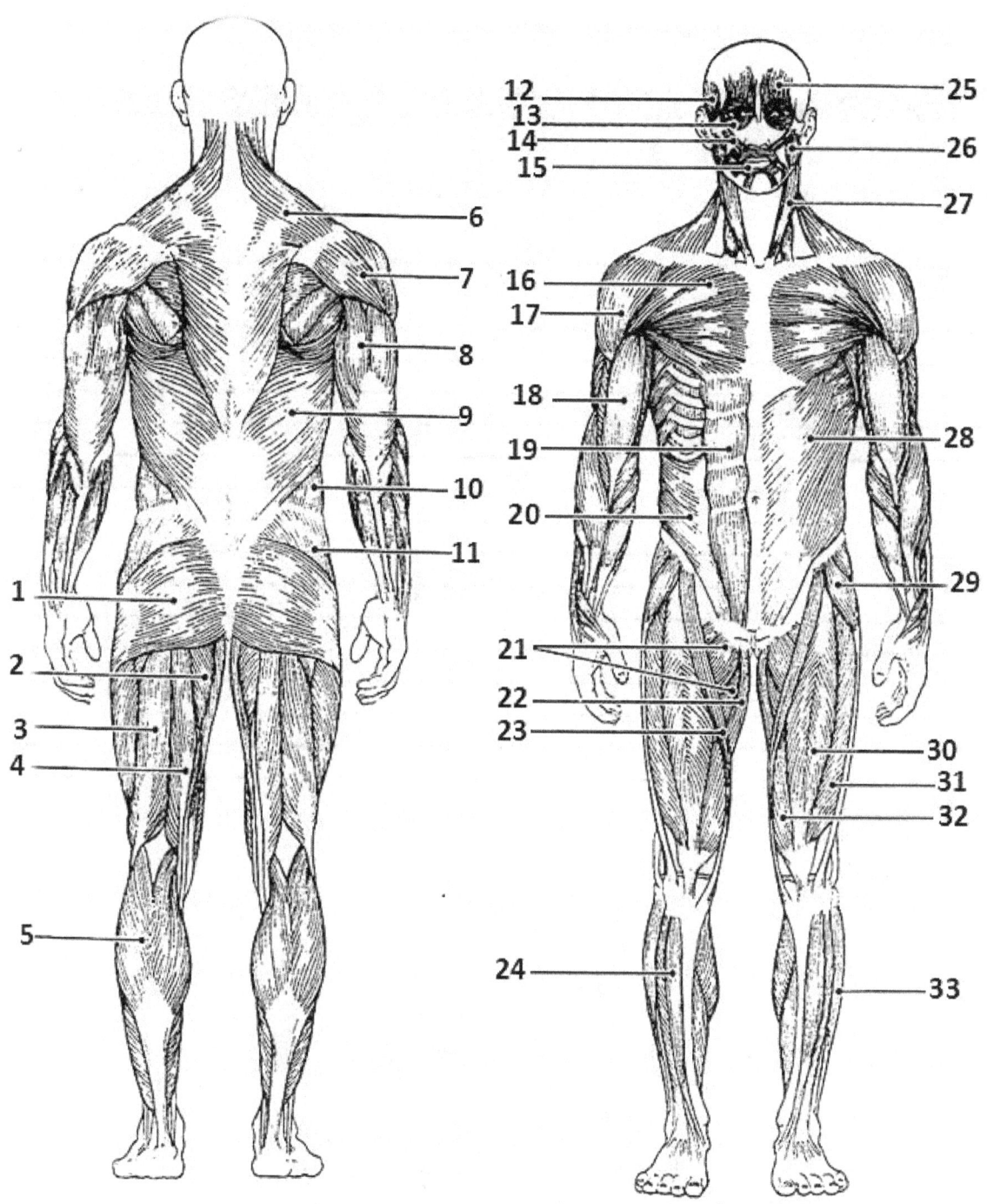

EXERCISE 8.5
COLORING SKELETAL MUSLCES

Purpose of exercise: To identify the names, origins, and insertions of facial muscles.

In this exercise you will color the facial muscles, name each muscle, and record the origin and insertion. Please remember to write the color used to identify each muscle.

Muscle	Origin	Insertion	Color
1.			
2.			
3.			
4.			
5.			
6.			
7.			
8.			
9.			
10.			
11.			
12.			
13.			
14.			
15.			

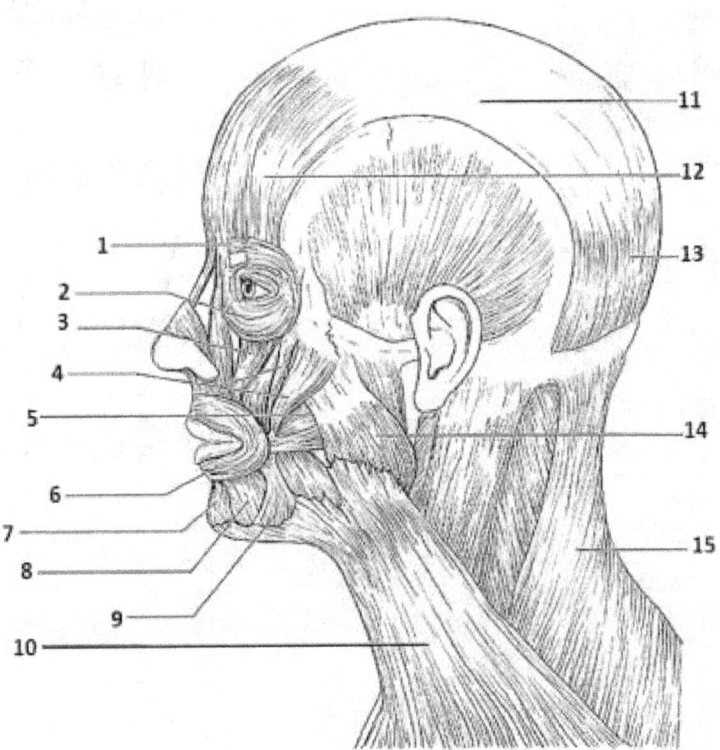

EXERCISE 8.6
COLORING SKELETAL MUSLCES

Purpose of exercise: To identify the names, origins, and insertions of superficial muscles of the shoulder girdle and upper arm.

In this exercise you will color superficial muscles of the shoulder girdle and upper arm, name each muscle, and record the origin and insertion. Please remember to write the color used to identify each muscle.

Muscle	Origin	Insertion	Color
1.			
2.			
3.			
4.			
5.			
6.			
7.			

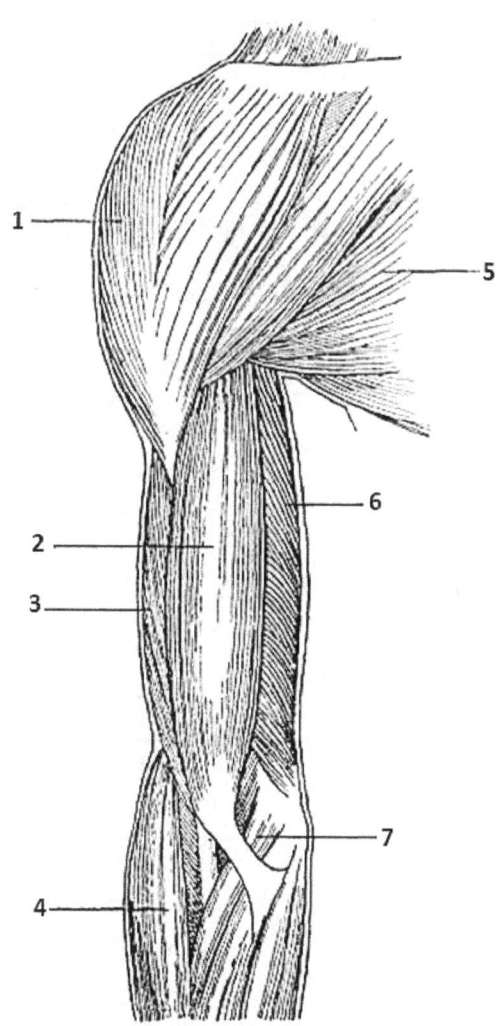

EXERCISE 8.7
COLORING SKELETAL MUSLCES

Purpose of exercise: To identify the names, origins, and insertions of deep muscles of the shoulder girdle and upper arm.

In this exercise you will color deep muscles of the shoulder girdle and upper arm, name each muscle, and record the origin and insertion. Please remember to write the color used to identify each muscle.

Muscle	Origin	Insertion	Color
1.			
2.			
3.			
4.			
5.			
6.			
7.			

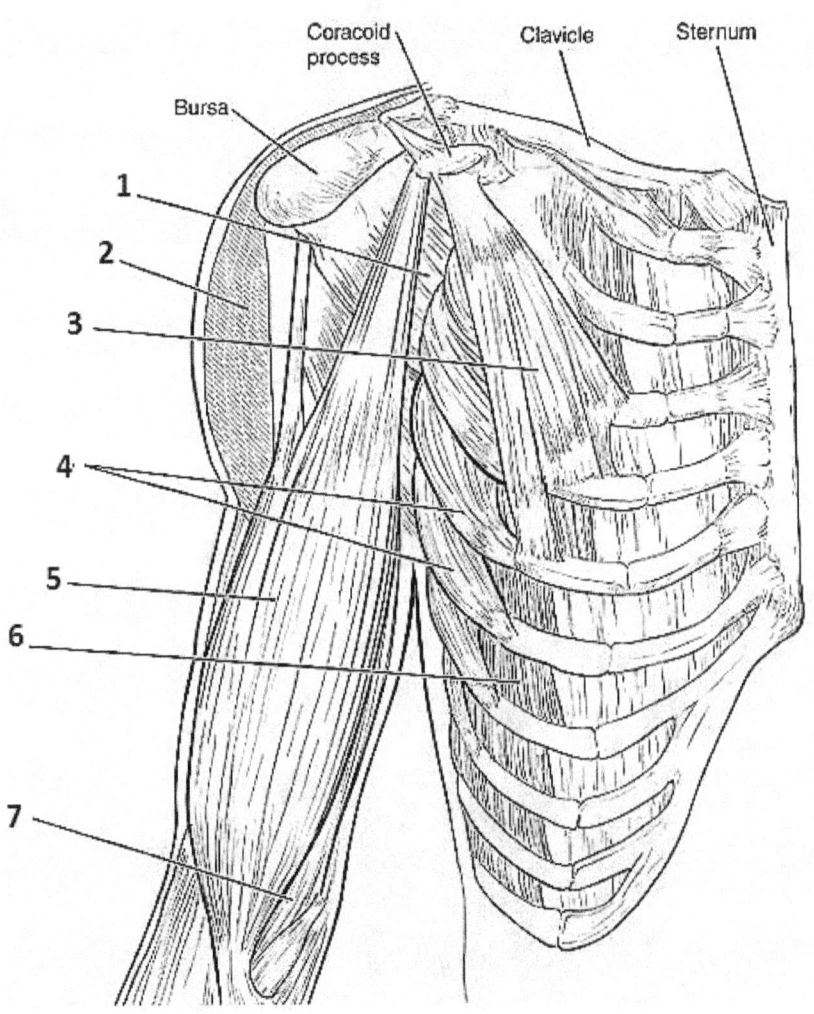

EXERCISE 8.8
COLORING SKELETAL MUSLCES

Purpose of exercise: To identify the names, origins, and insertions of superficial muscles of the forearm and hand.

In this exercise you will color superficial muscles of the forearm and hand, name each muscle, and record the origin and insertion. Please remember to write the color used to identify each muscle.

Muscle	Origin	Insertion	Color
1.			
2.			
3.			
4.			
5.			
6.			
7.			
8.			
9.			
10.			
11.			
12.			

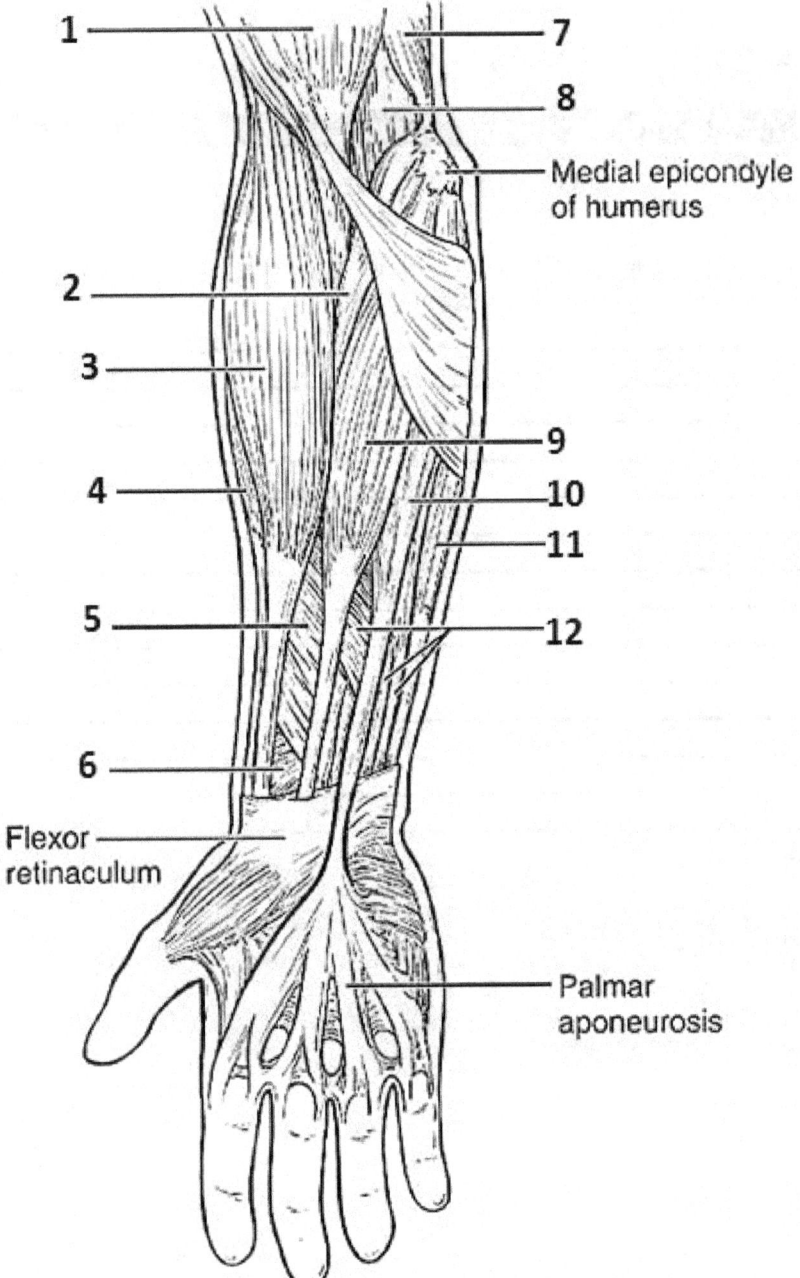

EXERCISE 8.9
COLORING SKELETAL MUSLCES

Purpose of exercise: To identify the names, origins, and insertions of superficial & deep muscles of the thorax.

In this exercise you will color superficial & deep muscles of the thorax, name each muscle, and record the origin and insertion. Please remember to write the color used to identify each muscle.

Muscle	Origin	Insertion	Color
1.			
2.			
3.			
4.			
5.			
6.			
7.			
8.			
9.			

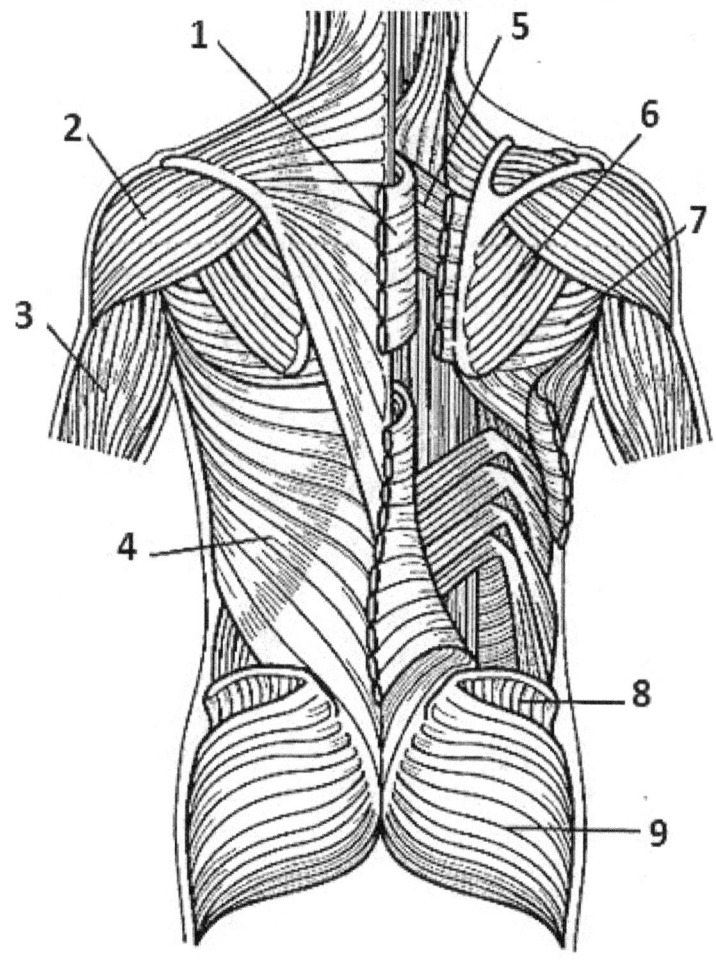

EXERCISE 8.10
COLORING SKELETAL MUSLCES

Purpose of exercise: To identify the names, origins, and insertions of muscles of the thigh.

In this exercise you will color muscles of the thigh, name each muscle, and record the origin and insertion. Please remember to write the color used to identify each muscle.

Muscle	Origin	Insertion	Color
1.			
2.			
3.			
4.			
5.			
6.			
7.			
8.			
9.			
10.			
11.			
12.			
13.			

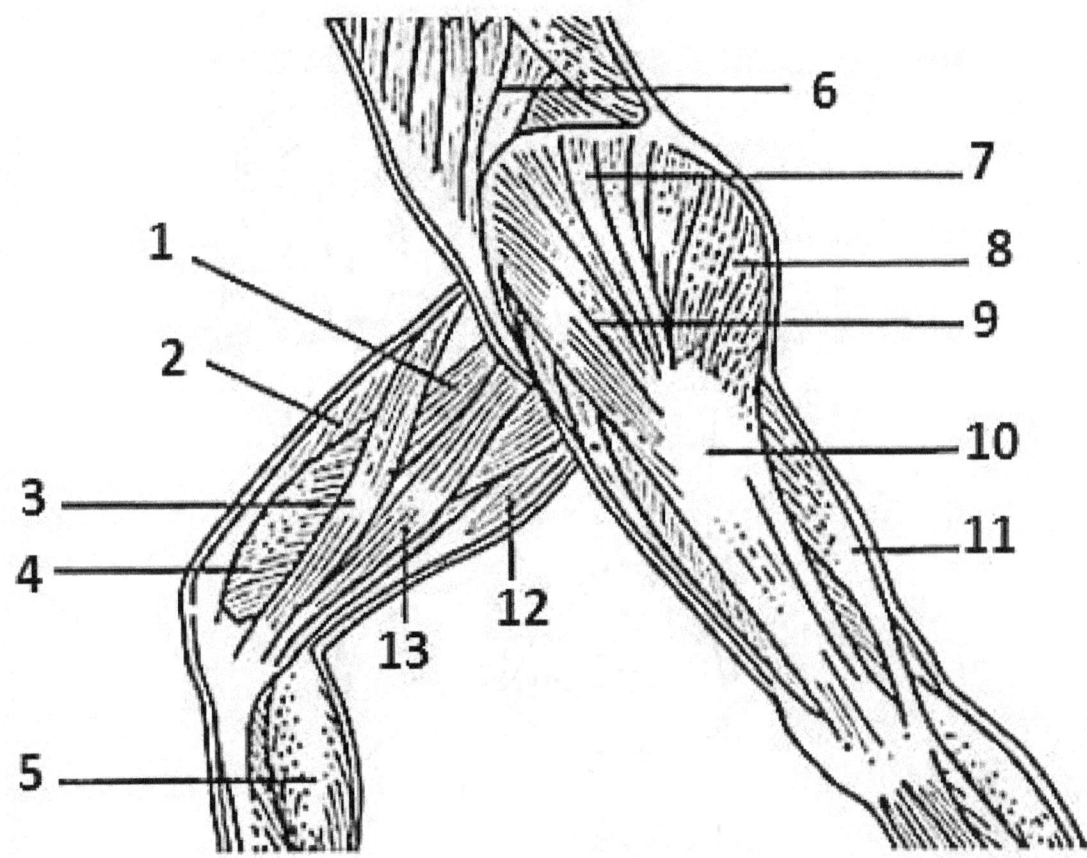

EXERCISE 8.11
COLORING SKELETAL MUSLCES

Purpose of exercise: To identify the names, origins, and insertions of muscles of the lower leg.

In this exercise you will color muscles of the lower leg, name each muscle, and record the origin and insertion. Please remember to write the color used to identify each muscle.

Muscle	Origin	Insertion	Color
1.			
2.			
3.			
4.			
5.			
6.			
7.			
8.			
9.			
10.			
11.			
12.			
13.			

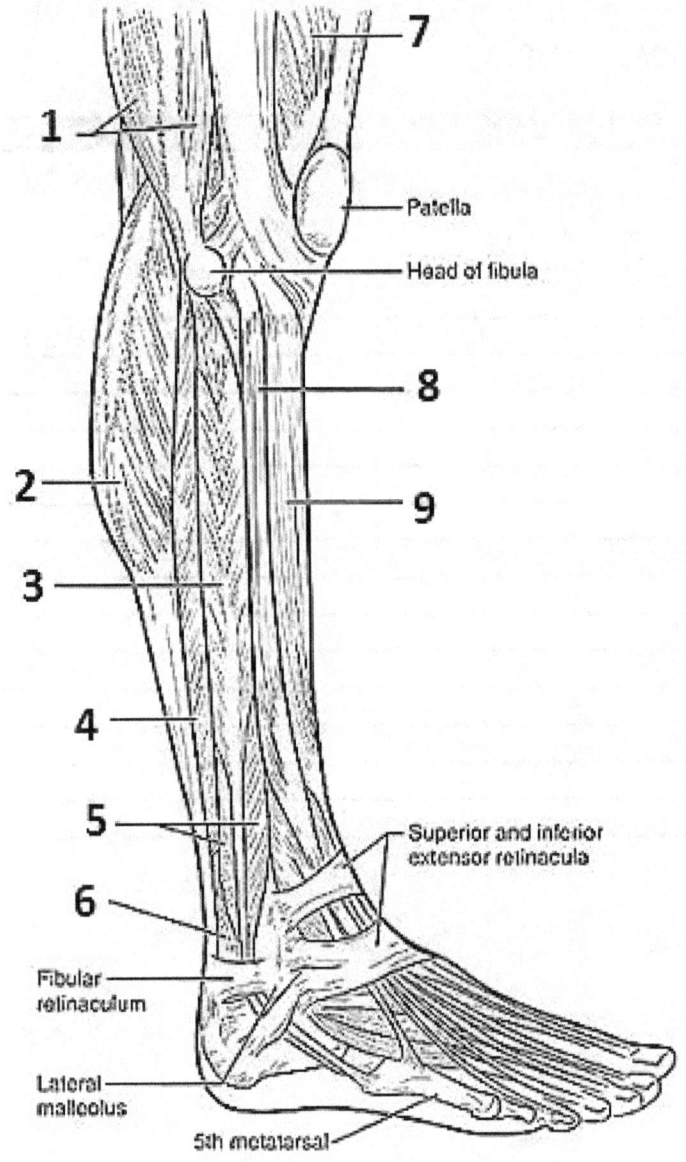

EXERCISE 8.12
MUSCLE RESPONSE TO ELECTRICAL STIMULATION

Purpose of exercise: To explore the relationship between workload and a muscle's threshold of stimulation.

Click on *Exercise 8.12* within your online platform or enter the address below into your web browser:
http://glencoe.mheducation.com/sites/dl/free/0078695104/383958/BL_21.html

EXERCISE 8.13
POKE–A–MUSCLE

Purpose of exercise: To help students learn the major superficial skeletal muscles of the human body.

Click on *Exercise 8.13* within your online platform or enter the address below into your web browser:
http://www.anatomyarcade.com/games/PAM/PAM.html

NERVOUS SYSTEM / SPECIAL SENSES
LAB 9

CRASHCOURSE VIDEO(S)

Click on the video embedded within your online platform or enter the address below into your web browser
1. https://youtu.be/qPix_X-9t7E
2. https://youtu.be/OZG8M_ldA1M
3. https://youtu.be/VitFvNvRIIY
4. https://youtu.be/q8NtmDrb_qo
5. https://youtu.be/QY9NTVh-Awo
6. https://youtu.be/71pCilo8k4M
7. https://youtu.be/0IDgB1CHVsA
8. https://youtu.be/qqU-VjqjczE
9. https://youtu.be/Ie2j7GpC4JU
10. https://youtu.be/o0DYP-u1rNM

(Please make sure to watch the video before continuing)

DEFINING KEY TERMS:

1. Acetylcholine (ACh):

2. Action potential:

3. All-or-None Response:

4. Astrocyte:

5. Axon:

6. Cerebellum:

7. Cerebrum:

8. Corpus callosum:

9. Dendrite:

10. Depolarization:

11. Dorsal root:

12. General (somatic) senses:

13. Gray Matter:

14. Hyperpolarization:

15. Hypothalamus:

16. Interneurons:

17. Medulla Oblongata:

18. Meninges:

19. Motor Neurons:

20. Neuroglia:

21. Neuron:

22. Pons:

23. Receptor:

24. Referred pain:

25. Reflex Arc:

26. Refractory Period:

27. Repolarization:

28. Resting Potential:

29. Sensation:

30. Sensory neurons:

31. Special senses:

32. Synapse:

33. Thalamus:

34. Ventral root:

35. White Matter:

EXERCISE 9.1
IDENTFYING PARTS OF NEURON

Purpose of exercise: To identify structural components of the neuron.

Click on *Exercise 9.1* within your online platform or enter the address below into your web browser:
http://learn.genetics.utah.edu/content/neuroscience/madneuron/

Now label the parts of the neuron.

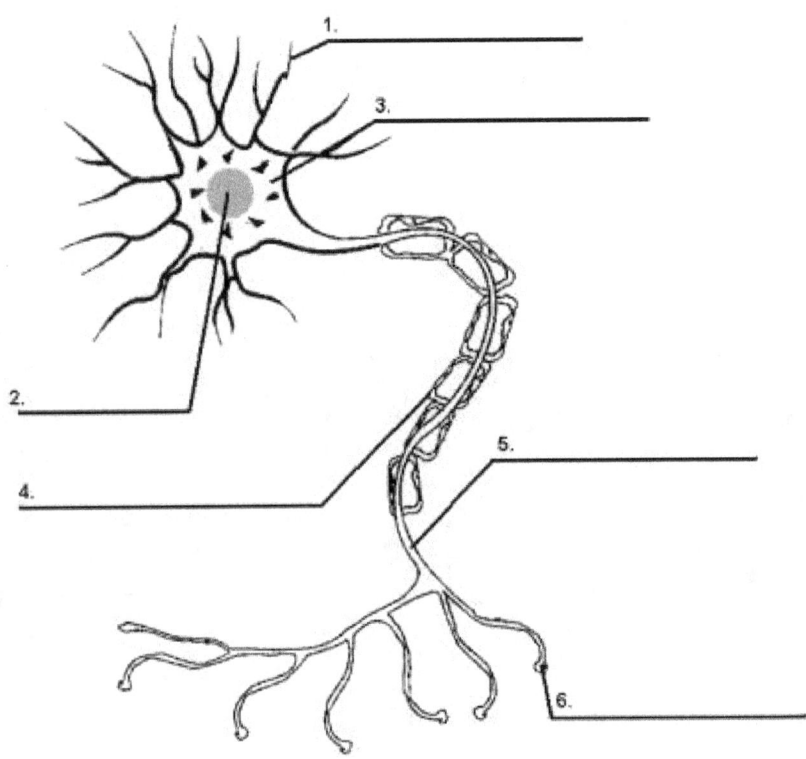

EXERCISE 9.2
MOLECULAR MECHANISM OF SYNAPTIC FUNCTION

Purpose of exercise: To demonstrate neuron synaptic transmission.

Click on *Exercise 9.2 (a)* within your online platform or enter the address below into your web browser:
http://learn.genetics.utah.edu/content/neuroscience/crossingdivide/

After you have used the previous hyperlink move to the next.

Click on *Exercise 9.2 (b)* within your online platform or enter the address below into your web browser:
http://www.hhmi.org/biointeractive/molecular-mechanism-synaptic-function

Please watch the 1 min 09 sec animation.

EXERCISE 9.3
NEURONS QUESTIONS

1. The brain and spinal cord form the _____ nervous system.
2. The cranial and spinal nerves form the _____ nervous system.
3. The nucleus of a neuron is found in its _____.
4. A(n) _____ of a neuron carries impulses toward the cell body.
5. A(n) _____ of a neuron carries impulses away from the cell body.
6. The _____ form the myelin sheath of peripheral neurons.
7. The _____ form the neurolemma of peripheral neurons.
8. The myelin sheath provides _____ for neurons.
9. The presence of a neurolemma may permit a peripheral neuron to _____.
10. The neuroglia that help form the blood–brain barrier are the _____.
11. The neuroglia that form the myelin sheath in the CNS are the _____.
12. The neuroglia that are capable of phagocytosis are the _____.
13. Sensory neurons carry impulses from _____ to the _____.
14. Motor neurons carry impulses from the _____ to _____.
15. Interneurons carry impulses within the _____.
16. A group of neurons with shared functions in the peripheral nervous system is called a(n) A group of neurons with shared functions in the CNS is called a(n) _____.
17. _____.
18. An electrical nerve impulse is carried by the _____ of a neuron.
19. A nerve impulse changes from electrical to chemical at a(n) _____.
20. Nerve impulses cross synapses by means of _____.
21. The electrical change called _____ takes place when a neuron receives a stimulus.
22. Depolarization means that _____ ions enter the neuron, and the outside of the membrane now has a(n) _____ charge.

EXERCISE 9.4
NEURONAL RESPONSE TO STIMULI

Purpose of exercise: To Identify the neurons based on the morphology and the response to stimuli, comparing them to previously published results. The student will record electrical activities of individual neurons while you deliver mechanical stimulus to the attached skin. Inject fluorescent dyes into the neurons to visualize their morphology.

Please note: The Full shockwave browser plug-in is required for this virtual lab. Download the latest version. Select "Full" when prompted for installer type.

Click on *Exercise 9.4* within your online platform or enter the address below into your web browser:
http://media.hhmi.org/biointeractive/vlabs/neurophysiology/index.html?_ga=2.111379715.602191367.1502809801-818520368.1502809801

Please follow the instructions of the website to complete the virtual lab.

EXERCISE 9.5
PARTS OF THE NERVOUS SYSTEM

Purpose of exercise: To identify structural components and divisions of the human nervous system.

Click on *Exercise 9.5(a)* within your online platform or enter the address below into your web browser:
https://www.getbodysmart.com/ap/nervoussystem/menu/menu.html

Please click each of the nine tutorials of the nervous system to further your understanding.

Now click on *Exercise 9.5(b)* within your online platform or enter the address below into your web browser:
https://www.biodigital.com/

- *(This is a free site, but you will need to sign up with your name and a validated email address. You can use your personal email address or school email address. MAKE SURE TO CREATE A PASSWORD YOU CAN REMEMBER. I WOULD SUGGEST WRITING IT DOWN AND KEEPING IT IN A SAFE PLACE. YOU WILL USE IT AGAIN.)*

Click **SIGN UP**. Once you have provided the appropriate information, you are now able to access BioDigital's website content. Click **LOG IN** using the email and password you created. Then click **SIGN IN**. On the left of your screen choose from the systems listed.

Please click on the square boxes next to this icon. [Search Human Library]. Now click on the **Anatomy By Systems** icon. Click on the following image(s) with caption. **Male Nervous System, Female Nervous System.** View the structures associated with this system.

EXERCISE 9.6
INTERACTIVE IDENTIFICATION OF THE BRAIN

Purpose of exercise: To identify parts of the nervous system using interactive tools.

Click on *Exercise 9.6 (a)* within your online platform or enter the address below into your web browser:
http://www.healthline.com/human-body-maps/brain

Please read and follow the instructions provided on the website. After you have used the previous hyperlink move to the next.

Click on *Exercise 9.6 (b)* within your online platform or enter the address below into your web browser:
https://www.nobelprize.org/educational/medicine/split-brain/splitbrainexp.html

EXERCISE 9.7
LABEL THE PARTS OF THE BRAIN

Purpose of exercise: To identify parts of the human brain.

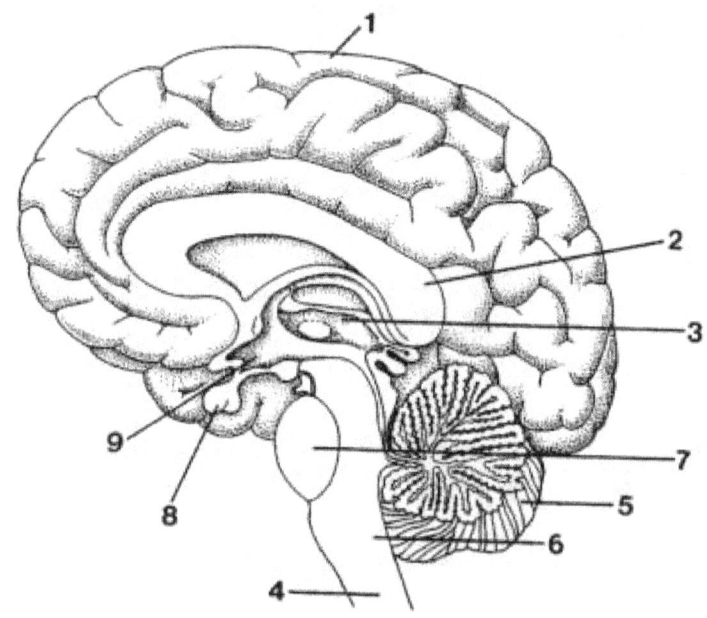

EXERCISE 9.8
LABEL THE LOBES OF THE BRAIN

Purpose of exercise: To identify lobes of the human cerebrum, brain stem, and cerebellum.

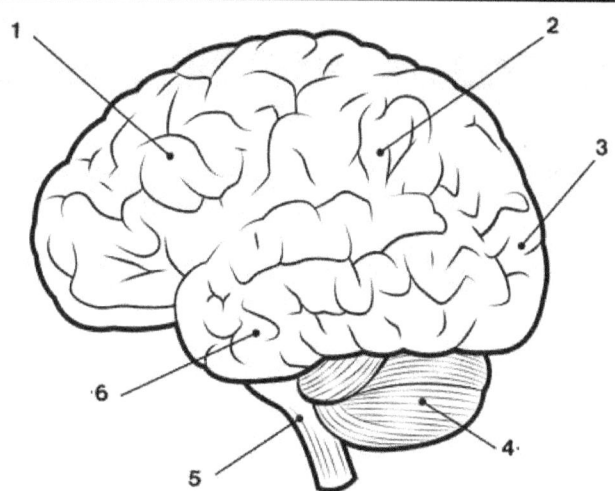

EXERCISE 9.9
LABEL THE PARTS OF THE SPINAL CORD

Purpose of exercise: To identify cross-sectional parts of the human spinal cord.

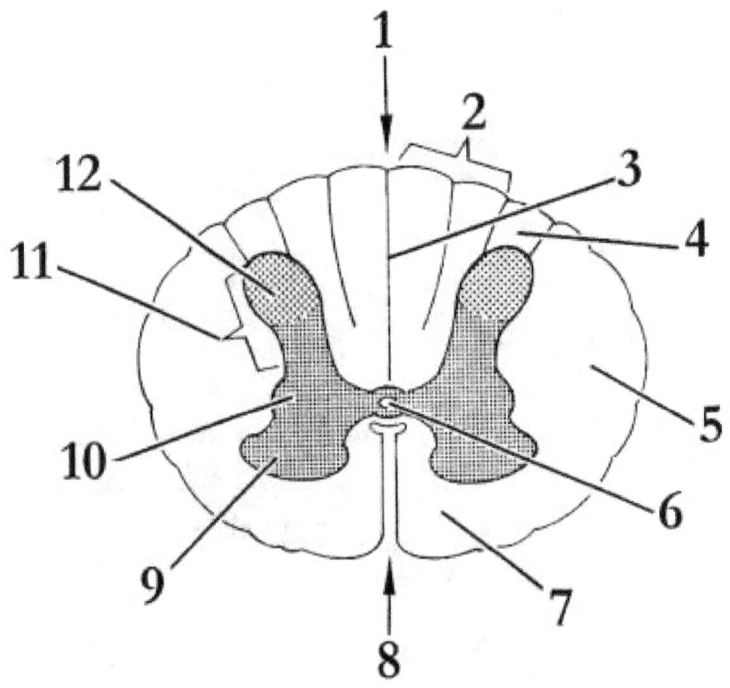

EXERCISE 9.10
SHEEP BRAIN DISSECTION

Purpose of exercise: To familiarize students with the three-dimensional structure of the brain.

Click on *Exercise 9.10* within your online platform or enter the address below into your web browser:
http://www.exploratorium.edu/memory/braindissection/index.html

Please click the **next** button on the page and read the instructions.

EXERCISE 9.11
CRANIAL NERVE MNEMONIC

Purpose of exercise: To create a mnemonic to assist in the identification of the twelve pairs of cranial nerves.

A mnemonic is an instructional strategy designed to help students improve their memory of important information. This technique connects new learning to prior knowledge through the use of visual and/or acoustic cues. The basic types of mnemonic strategies rely on the use of key words, rhyming words, or acronyms.

In this exercise, you will come up with your own mnemonic for the twelve pairs of cranial nerves.

_____

_____

_____

_____

_____

_____

EXERCISE 9.12
NEUROLOGICAL EXAMINATION

Purpose of exercise: To demonstrate the procedures of a routine neurologic examination.

A neurological examination, also called a neuro exam, is an evaluation of a person's nervous system that can be performed in the physician's office. It may be performed with instruments, such as lights and reflex hammers, and usually does not cause any pain to the patient. The nervous system consists of the brain, the spinal cord, and the nerves from these areas. There are many aspects of this examination, including an assessment of motor and sensory skills, balance and coordination, mental status (the patient's level of awareness and interaction with the environment), reflexes, and functioning of the nerves. The extent of the examination depends on many factors, including the initial problem that the patient is experiencing, the age of the patient, and the condition of the patient.

During a neurological examination, the physician will "test" the functioning of the nervous system. The nervous system is very complex and controls many parts of the body. The nervous system consists of the brain, spinal cord, 12 nerves that come from the brain, and the nerves that come from the spinal cord.

Materials
- Reflex Hammer
- C512 Tuning Fork
- Pen Light
- Cotton Swab

General Considerations
- Always consider left to right symmetry
- Consider central vs. peripheral deficits
- Organize your thinking into seven categories:
    1. Mental Status
    2. Cranial Nerves
    3. Motor
    4. Coordination and Gait
    5. Reflexes
    6. Sensory
    7. Special Tests

Cranial Number:	Cranial Name:	Major Functions:
I	Olfactory	Smell
II	Optic	Vision
III	Oculomotor	Eyelid and eyeball movement
IV	Trochlear	Innervates superior oblique, turns eye downward and laterally
V	Trigeminal	Chewing, face & mouth touch & pain
VI	Abducens	Turns eye laterally
VII	Facial	Controls most facial expressions, secretion of tears & saliva, taste
VIII	Vestibulocochlear aka (Auditory)	Hearing, equilibrium sensation
IX	Glossopharyngeal	Taste, senses carotid blood pressure
X	Vagus	Senses aortic blood pressure, slows heart rate, stimulates digestive organs, taste
XI	Spinal Accessory	Controls trapezius & sternocleidomastoid, controls swallowing movements
XII	Hypoglossal	Controls tongue movements

Observation
- Ptosis (III)
- Facial Droop or Asymmetry (VII)
- Hoarse Voice (X)
- Articulation of Words (V, VII, X, XII)
- Abnormal Eye Position (III, IV, VI)
- Abnormal or Asymmetrical Pupils (II, III)

Mental Status
The <u>Mini Mental Status Examination</u> is a useful screening tool. The Mini Mental State Examination (MMSE) is a tool that can be used to systematically and thoroughly assess mental status. It is an 11-question measure that tests

five areas of cognitive function: orientation, registration, attention and calculation, recall, and language. The maximum score is 30. A score of 23 or lower is indicative of cognitive impairment. The MMSE takes only 5-10 minutes to administer and is therefore practical to use repeatedly and routinely.

The Mini-Mental Status Exam

Patient _____ Examiner _____ Date_____

Maximum	Score	
5	○ 0 out of 5 (0 points) ○ 1 out of 5 (1 points) ○ 2 out of 5 (2 points) ○ 3 out of 5 (3 points) ○ 4 out of 5 (4 points) ○ 5 out of 5 (5 points)	What is the (year) (season) (date) (day) (month)?
5	○ 0 out of 5 (0 points) ○ 1 out of 5 (1 points) ○ 2 out of 5 (2 points) ○ 3 out of 5 (3 points) ○ 4 out of 5 (4 points) ○ 5 out of 5 (5 points)	Ask the patient "Where are we (state) (country) (town) (floor)?
3	○ Says 0 out of 3 (0 points) ○ Says 1 out of 3 (1 points) ○ Says 2 out of 3 (2 points) ○ Says 3 out of 3 (3 points)	Tell patient "I'd like to test your memory; Examiner will tell patient to repeat these words after they've said all three: boat, cucumber, wire
5	○ Cannot do it at all (0 points) ○ Gets 1 right (1 points) ○ Gets 2 right (2 points)	Tell patient "Begin with 100 and count backwards by 5" Answers = (100, 95, 90, 85, 80, 75, etc.)

	○ Gets 3 right (3 points) ○ Gets 4 right (4 points) ○ Gets 5 right (5 points)	
3	○ Gets 0 out of 3 (0 points) ○ Gets 1 out of 3 (1 points) ○ Gets 2 out of 3 (2 points) ○ Gets all 3 (3 points)	Tell patient "Can you name the three objects I named before?"
2	○ Gets neither one right (0 points) ○ Gets 1 of 2 right (1 points) ○ Gets them both right (2 points)	Place several items in front of the patient and tell them to "Name these items" Watch and see how they respond
1	○ Does not say it properly (0 points) ○ Says it properly (1 points)	Tell patient to repeat the following: "No ifs, ands, or buts."
3	○ Does none of these 3 things (0 points) ○ Does 1 of these 3 things (1 points) ○ Does 2 of these 3 things (2 points) ○ Does 3 of these 3 things (3 points)	Ask the patient to, "Take a sheet of paper in their right hand, fold it in half, and put it on the floor."
1	○ Patient does not close eyes (0 points) ○ Patient closes eyes (1 points)	Ask the patient to read and obey the following. The Examiner will write on a sheet of paper, "CLOSE YOUR EYES".
1	○ Patient does not write a sentence (0 points) ○ Patient writes a sentence (1 points)	Ask the patient to write a sentence.
1	○ Patient does not copy the design properly (0 points) ○ Patient copies the design properly (1 points)	The Examiner will draw interlocking triangles, then they should ask the patient to draw a copy of their design.

30	Total pts._____	

Testing Cranial Nerves

I – Olfaction

Click on *Exercise 9.12 (a)* within your online platform or enter the address below into your web browser:
http://www.neuroexam.com/neuroexam/content16.html

II - Optic

Test Visual Acuity

Click on *Exercise 9.12 (b)* within your online platform or enter the address below into your web browser. Read the information and follow the direction.
http://www.personaleyes.com.au/online-eye-test/index.php

Screen Visual Fields by Confrontation
0. Stand two feet in front of the patient and have them look into your eyes.
1. Hold your hands about one foot away from the patient's ears, and wiggle a finger on one hand.
2. Ask the patient to indicate which side they see the finger move.
3. Repeat two or three times to test both temporal fields.
4. If an abnormality is suspected, test the four quadrants of each eye while asking the patient to cover the opposite eye with a card.

Pupillary Responses (CN II, III)

Click on *Exercise 9.12 (c)* within your online platform or enter the address below into your web browser. Read the information and watch the video before performing the pupillary response assessment.
http://www.neuroexam.com/neuroexam/content19.html

Test Pupillary Reactions to Light
1. Dim the room lights as necessary.
2. Ask the patient to look into the distance.
3. Shine a bright light obliquely into each pupil.
4. Look for both the **direct** (same eye) and **consensual** (other eye) reactions.
5. Watch the patient's pupil size for asymmetry or irregularity.

Test Pupillary Reactions to Accommodation
1. Hold your finger about 10cm from the patient's nose.
2. Ask them to alternate looking into the distance and at your finger.
3. Observe the pupillary response in each eye.

III - Oculomotor
- Observe for Ptosis (when the upper eyelid droops over the eye)
- **Test Extraocular Movements**

1. Stand or sit 3 to 6 feet in front of the patient.
2. Ask the patient to follow your finger with their eyes without moving their head.
3. Check gaze in the six cardinal directions using a cross or "H" pattern.
4. Pause during upward and lateral gaze to check for nystagmus.
5. Check convergence by moving your finger toward the bridge of the patient's nose.

IV - Trochlear
Test Extraocular Movements (Inward and Down Movement)

V - Trigeminal
- **Test Temporal and Masseter Muscle Strength**
    1. Ask patient to both open their mouth and clench their teeth.
    2. Palpate the temporal and massetter muscles as they do this.
- **Test the Three Divisions for Pain Sensation**
    1. Explain what you intend to do.
    2. Use a suitable sharp object to test the forehead, cheeks, and jaw on both sides.
    3. Substitute a blunt object occasionally and ask the patient to report "sharp" or "dull."

VI - Abducens
Test Extraocular Movements (Lateral Movement)

VII - Facial
- Observe for Any Facial Droop or Asymmetry
- Ask Patient to do the following, note any lag, weakness, or assymetry:
    1. Raise eyebrows
    2. Close both eyes to resistance
    3. Smile
    4. Frown
    5. Show teeth
    6. Puff out cheeks

VIII - Acoustic
- **Screen Hearing**
    1. Face the patient and hold out your arms with your fingers near each ear.
    2. Rub your fingers together on one side while moving the fingers noiselessly on the other.
    3. Ask the patient to tell you when and on which side they hear the rubbing.
    4. Increase intensity as needed and note any asymmetry.
    5. If abnormal, proceed with the Weber and Rinne tests.
- **Test for Lateralization (Weber)**
    1. Use a tuning fork.
    2. Start the fork vibrating by tapping it on your opposite hand.
    3. Place the base of the tuning fork firmly on top of the patient's head.
    4. Ask the patient where the sound appears to be coming from (normally in the midline).
- **Compare Air and Bone Conduction (Rinne)**
    1. Use a tuning fork.
    2. Start the fork vibrating by tapping it on your opposite hand.
    3. Place the base of the tuning fork against the mastoid bone behind the ear.
    4. When the patient no longer hears the sound, hold the end of the fork near the patient's ear (air conduction is normally greater than bone conduction).

X - Vagus
1. Listen to the patient's voice, is it hoarse or nasal?
2. Ask Patient to Swallow

3. Ask Patient to Say "Ah"
    a. Watch the movements of the soft palate and the pharynx.

XI - Accessory
1. From behind, look for atrophy or asymmetry of the trapezius muscles.
2. Ask patient to shrug shoulders against resistance.
3. Ask patient to turn their head against resistance. Watch and palpate the sternocleidomastoid muscle on the opposite side.

XII - Hypoglossal
1. Listen to the articulation of the patient's words.
2. Observe the tongue as it lies in the mouth
3. Ask patient to:
    a. Protrude tongue
    b. Move tongue from side to side

Motor
1. **Observation**
   - Involuntary Movements
   - Muscle Symmetry
     - Left to Right
     - Proximal vs. Distal
   - Atrophy
     - Pay particular attention to the hands, shoulders, and thighs.
   - Gait (Watch patient as they walk)

2. **Muscle Tone**
Click on *Exercise 9.12 (d)* within your online platform or enter the address below into your web browser. Read the information and watch the video before performing the muscle tone assessment.
http://www.neuroexam.com/neuroexam/content28.html#tone

   1. Ask the patient to relax.
   2. Flex and extend the patient's fingers, wrist, and elbow.
   3. Flex and extend patient's ankle and knee.
   4. There is normally a small, continuous resistance to passive movement.
   5. Observe for decreased (flaccid) or increased (rigid/spastic) tone.

3. **Muscle Strength**
Click on *Exercise 9.12 (e)* within your online platform or enter the address below into your web browser. Read the information and watch the video before performing the muscle strength assessment.
http://www.neuroexam.com/neuroexam/content29.html

   - Test strength by having the patient move against your resistance.
   - Always compare one side to the other.
   - Grade strength on a scale from 0 to 5 "out of five":

Grading Motor Strength	
Grade	Description
0/5	No muscle movement
1/5	Visible muscle movement, but no movement at the joint
2/5	Movement at the joint, but not against gravity
3/5	Movement against gravity, but not against added resistance
4/5	Movement against resistance, but less than normal
5/5	Normal strength

- Test the following:
    1. Flexion at the elbow (C5, C6, biceps)
    2. Extension at the elbow (C6, C7, C8, triceps)
    3. Extension at the wrist (C6, C7, C8, radial nerve)
    4. Squeeze two of your fingers as hard as possible ("grip," C7, C8, T1)
    5. Finger abduction (C8, T1, ulnar nerve)
    6. Opposition of the thumb (C8, T1, median nerve)
    7. Flexion at the hip (L2, L3, L4, iliopsoas)
    8. Adduction at the hips (L2, L3, L4, adductors)
    9. Abduction at the hips (L4, L5, S1, gluteus medius and minimus)
    10. Extension at the hips (S1, gluteus maximus)
    11. Extension at the knee (L2, L3, L4, quadriceps)
    12. Flexion at the knee (L4, L5, S1, S2, hamstrings)
    13. Dorsiflexion at the ankle (L4, L5)
    14. Plantar flexion (S1)

4. **Pronator Drift**
    1. Ask the patient to stand for 20-30 seconds with both arms straight forward, palms up, and eyes closed.
    2. Instruct the patient to keep the arms still while you tap them briskly downward.
    3. The patient will not be able to maintain extension and supination (and "drift into pronation) with upper motor neuron disease.

Coordination and Gait
Click on *Exercise 9.12 (f)* within your online platform or enter the address below into your web browser. Read the information on the page and watch the video before performing the assessment below.
http://www.neuroexam.com/neuroexam/content36.html

1. **Rapid Alternating Movements**
    1. Ask the patient to strike one hand on the thigh, raise the hand, turn it over, and then strike it back down as fast as possible.
    2. Ask the patient to tap the distal thumb with the tip of the index finger as fast as possible.
    3. Ask the patient to tap your hand with the ball of each foot as fast as possible.

2. **Point-to-Point Movements**
    1. Ask the patient to touch your index finger and their nose alternately several times. Move your finger about as the patient performs this task.

2. Hold your finger still so that the patient can touch it with one arm and finger outstretched. Ask the patient to move their arm and return to your finger with their eyes closed.
3. Ask the patient to place one heel on the opposite knee and run it down the shin to the big toe. Repeat with the patient's eyes closed.

3. **Romberg**

    Click on *Exercise 9.12 (g)* within your online platform or enter the address below into your web browser. Read the information and watch the video before performing the Romberg Test.
    http://www.neuroexam.com/neuroexam/content37.html

    1. Be prepared to catch the patient if they are unstable.
    2. Ask the patient to stand with the feet together and eyes closed for 5-10 seconds without support.
    3. The test is said to be positive if the patient becomes unstable (indicating a vestibular or proprioceptive problem).

4. **Gait**

    Ask the patient to:
    1. Walk across the room, turn and come back
    2. Walk heel-to-toe in a straight line
    3. Walk on their toes in a straight line
    4. Walk on their heels in a straight line
    5. Hop in place on each foot
    6. Do a shallow knee bend
    7. Rise from a sitting position

Reflexes

1. **Deep Tendon Reflexes**

    Click on *Exercise 9.12 (h)* within your online platform or enter the address below into your web browser. Read the information and watch the video before performing the Deep Tendon Reflex assessment.
    http://www.neuroexam.com/neuroexam/content31.html

    1. The patient must be relaxed and positioned properly before starting.
    2. Reflex response depends on the force of your stimulus. Use no more force than you need to provoke a definite response.
    3. Reflexes can be reinforced by having the patient perform isometric contraction of other muscles (clenched teeth).
    4. Reflexes should be graded on a 0 to 4 "plus" scale:

Tendon Reflex Grading Scale	
Grade	Description
0	Absent
1+ or +	Hypoactive
2+ or ++	"Normal"
3+ or +++	Hyperactive without clonus
4+ or ++++	Hyperactive with sustained clonus

- Biceps (C5, C6)
    1. The patient's arm should be partially flexed at the elbow with the palm down.
    2. Place your thumb or finger firmly on the biceps tendon.
    3. Strike your finger with the reflex hammer.
    4. You should feel the response even if you can't see it.
- Triceps (C6, C7)
    1. Support the upper arm and let the patient's forearm hang free.
    2. Strike the triceps tendon above the elbow with the broad side of the hammer.
    3. If the patient is sitting or lying down, flex the patient's arm at the elbow and hold it close to the chest.
- Brachioradialis (C5, C6)
    1. Have the patient rest the forearm on the abdomen or lap.
    2. Strike the radius about 1-2 inches above the wrist.
    3. Watch for flexion and supination of the forearm.
- Abdominal (T8, T9, T10, T11, T12)
    1. Use a blunt object such as a key or tongue blade.
    2. Stroke the abdomen lightly on each side in an inward and downward direction above (T8, T9, T10) and below the umbilicus (T10, T11, T12).
    3. Note the contraction of the abdominal muscles and deviation of the umbilicus towards the stimulus.
- Knee (L2, L3, L4)
    1. Have the patient sit or lie down with the knee flexed.
    2. Strike the patellar tendon just below the patella.
    3. Note contraction of the quadriceps and extension of the knee.
- Ankle (S1, S2)
    1. Dorsiflex the foot at the ankle.
    2. Strike the Achilles tendon.
    3. Watch and feel for plantar flexion at the ankle.

Clonus ------ (Clonus is involuntary and rhythmic muscle contractions caused by a permanent lesion in descending motor neurons. Clonus tends to co-exist with spasticity in many cases of stroke and spinal cord injury likely due to their common physiological origins)

If the reflexes seem hyperactive, test for ankle clonus:
1. Support the knee in a partly flexed position.
2. With the patient relaxed, quickly dorsiflex the foot.
3. Observe for rhythmic oscillations.

2. **Plantar Response (Babinski)**

   1. Stroke the lateral aspect of the sole of each foot with the end of a reflex hammer or key.
   2. Note movement of the toes, normally flexion (withdrawal).
   3. Extension of the big toe with fanning of the other toes is abnormal. This is referred to as a positive Babinski.

Sensory

1. **General**
    1. Explain each test before you do it.
    2. Unless otherwise specified, the patient's eyes should be closed during the actual testing.
    3. Compare symmetrical areas on the two sides of the body.
    4. Also compare distal and proximal areas of the extremities.
    5. When you detect an area of sensory loss map out its boundaries in detail.

2. **Vibration**
   - Use tuning fork.
       1. Test with a non-vibrating tuning fork first to ensure that the patient is responding to the correct stimulus.
       2. Place the stem of the fork over the distal interphalangeal joint of the patient's index fingers and big toes.
       3. Ask the patient to tell you if they feel the vibration.

   - If vibration sense is impaired proceed proximally:
       1. Wrists
       2. Elbows
       3. Medial malleoli
       4. Patellas
       5. Anterior superior iliac spines
       6. Spinous processes
       7. Clavicles

3. **Subjective Light Touch**
   1. Use your fingers to touch the skin lightly on both sides simultaneously.
   2. Test several areas on both the upper and lower extremities.
   3. Ask the patient to tell you if there is difference from side to side or other "strange" sensations.

4. **Position Sense**
   1. Grasp the patient's big toe and hold it away from the other toes to avoid friction.
   2. Show the patient "up" and "down."
   3. With the patient's eyes closed ask the patient to identify the direction you move the toe.
   4. If position sense is impaired move proximally to test the ankle joint.
   5. Test the fingers in a similar fashion.
   6. If indicated move proximally to the metacarpophalangeal joints, wrists, and elbows.

EXERCISE 9.13
EYE JIGSAW PUZZLE

Purpose of exercise: To identify parts of the human eye using a virtual interactive jigsaw puzzle.

Click on *Exercise 9.13* within your online platform or enter the address below into your web browser. Read the information and watch the video.
http://www.anatomyarcade.com/games/jigsaws/EyeJigsaw/eyeJigsaw.html

Please read and follow the instructions provided on the website.

EXERCISE 9.14
IDENTFYING PARTS OF THE EYE

Purpose of exercise: To identify parts of the human eye.

Now click on *Exercise 9.14* within your online platform or enter the address below into your web browser:
https://www.biodigital.com/

- *(This is a free site, but you will need to sign up with your name and a validated email address. You can use your personal email address or school email address. MAKE SURE TO CREATE A PASSWORD YOU CAN REMEMBER. I WOULD SUGGEST WRITING IT DOWN AND KEEPING IT IN A SAFE PLACE. YOU WILL USE IT AGAIN.)*

Click **SIGN UP**. Once you have provided the appropriate information, you are now able to access BioDigital's website content. Click **LOG IN** using the email and password you created. Then click **SIGN IN**. On the left of your screen choose from the systems listed.

Please click on the square boxes next to this icon. . Now click on the **Anatomy By Systems** icon. Click on the following image(s) with caption. **Male Visual System, Female Visual System.** View the structures associated with this system.

Label the parts of the human eye.

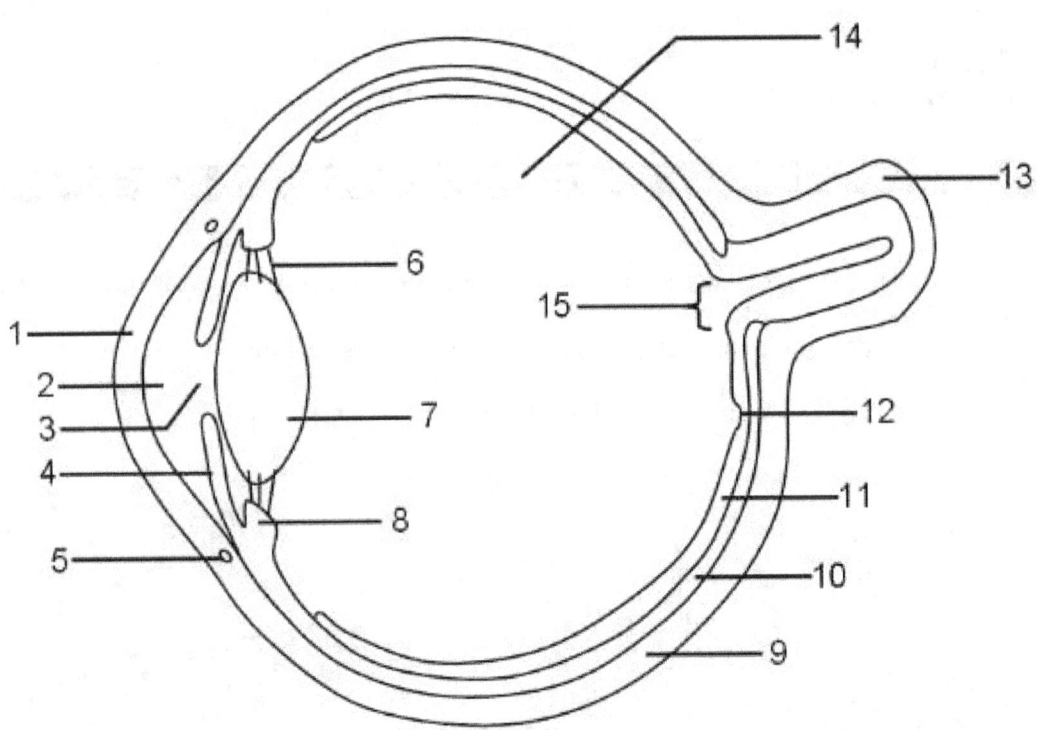

EXERCISE 9.15
COLOR VISION

Purpose of exercise: To determine what color the person sees for various combinations of light.

Click on *Exercise 9.15* within your online platform or enter the address below into your web browser.
https://phet.colorado.edu/sims/html/color-vision/latest/color-vision_en.html

Please read and follow the instructions provided on the website.

EXERCISE 9.16
VIRTUAL SHEEP EYE DISSECTION

Purpose of exercise: To familiarize students with the three-dimensional structure of the eye.

Click on *Exercise 9.16* within your online platform or enter the address below into your web browser.
http://science.jburroughs.org/resources/skeleton/eye/eyedissection.html

Please read and follow the instructions provided on the website. "**Click here to continue the dissection**" link on the website to move through the virtual sheep eye dissection.

EXERCISE 9.17
SURGERIES OF THE EYE

Purpose of exercise: To explore various type of surgeries associated with the human eye.

Please make sure that your Adobe Flash Player is updated on your computer. Also, please be care not to click on any of the third party advertisements because they will route you to another site.
You will perform the following virtual surgeries: **Retinal Detachment Eye Surgery, Cataract Eye Surgery,** and **LASIK Eye Surgery.**

Click on *Exercise 9.17 (a)* within your online platform or enter the address below into your web browser.
Retinal Detachment Eye Surgery: http://www.surgerysquad.com/surgeries/detached-retina-surgery/

Click on *Exercise 9.17 (b)* within your online platform or enter the address below into your web browser.
Cataract Eye Surgery: http://www.surgerysquad.com/surgeries/cataract-eye-surgery/

Click on *Exercise 9.17 (c)* within your online platform or enter the address below into your web browser.
LASIK Eye Surgery: http://www.surgerysquad.com/surgeries/virtual-lasik-eye-surgery/

Once you are ready to begin the surgery, click **START**. Please listen and read the instructions provided on the website and follow the interactive steps to perform the various surgeries.

EXERCISE 9.18
UNDERSTANDING THE EYES

Purpose of exercise: To familiarize students' with conditions associated with the human eye.

Click on *Exercise 9.18* within your online platform or enter the address below into your web browser.
https://share.rendia.com/theater/e500fc92-4827-11e5-97c3-22000a8906f4

Please watch videos 1-6 on this page.

EXERCISE 9.19
THE EAR GAME

Purpose of exercise: To familiarize students' with the functions and parts of the human ear.

Click on *Exercise 9.19* within your online platform or enter the address below into your web browser.
https://www.nobelprize.org/educational/medicine/ear/game/index.html

Please listen and follow the instructions provided on the website to play the game.

EXERCISE 9.20
SLEEP STUDY

Purpose of exercise: To familiarize students' with the function and parts of the human ear.

Polysomnography, also called a sleep study, is a test used to diagnose sleep disorders. Polysomnography records your brain waves, the oxygen level in your blood, heart rate and breathing, as well as eye and leg movements during the study. Polysomnography usually is done at a sleep disorders unit within a hospital or at a sleep center. You'll be asked to come to the sleep center in the evening for polysomnography so that the test can record your nighttime sleep patterns. Polysomnography is occasionally done during the day to accommodate shift workers who habitually sleep during the day.

In addition to helping diagnose sleep disorders, polysomnography may be used to help adjust your treatment plan if you've already been diagnosed with a sleep disorder. Polysomnography monitors your sleep stages and cycles to identify if or when your sleep patterns are disrupted and why. The normal process of falling asleep begins with a sleep stage called non-rapid eye movement (NREM) sleep. During this stage, your brain waves, as recorded by electroencephalography (EEG), slow down considerably. Your eyes don't move back and forth rapidly during NREM, in contrast to later stages of sleep. After an hour or two of NREM sleep, your brain activity picks up again, and rapid eye movement (REM) sleep begins. Most dreaming occurs during REM sleep.

You normally go through four to six sleep cycles a night, cycling between NREM and REM sleep in about 90 minutes. Your REM stage usually lengthens with each cycle as the night progresses. Sleep disorders can disturb this sleep process. Polysomnography monitors your sleep stages and cycles to identify if or when your sleep patterns are disrupted. Your doctor may recommend polysomnography if he or she suspects you have: Sleep apnea or another sleep-related breathing disorder, Periodic limb movement disorder, Narcolepsy, REM sleep behavior disorder, Unusual behaviors during sleep, and unexplained chronic insomnia.

Sleep studies are expensive, time consuming, and often unavailable in many locations. However, there are simple tests you can perform at home that will give you a general idea of how these studies are implemented and an appreciation for Polysomnographic Technologists and practitioners.

ASSESSING RAPID EYE MOVEMENT SLEEP (REM)

During REM sleep, the eyes move back and forth. There are roughly 4-5 REM periods in a night's sleep. Even though most people eyes are closed (or partially closed) when they are a sleep, you can still detect movement of their eyeballs through their eyelids.

Practice observing this in someone who is awake. Just ask the person to close his or her eyes and then to move their eyes. You should be able to see a bulge moving behind the eyelid. Now you are ready conduct our exercise.

Procedure
1. When your subject (father, mother, child, husband, wife, etc.) is sleeping, take a peek at their eyes.
2. Are their eyes moving back and forth rapidly?
3. If they are, the person is probably in REM sleep.

HYPNAGOGIA

Just before you fall asleep, you go into a special state of mind that scientists call hynogogia. Hypnagogia is the experience of the transitional state from wakefulness to sleep: the hypnagogic state of consciousness, during the onset of sleep. Mental phenomena that occur during this "threshold consciousness" phase include lucid thought, lucid dreaming, hallucinations, and sleep paralysis. During sleep paralysis one is aware but unable to move.

Unfortunately, most people aren't able to focus their attention on hypnogogic experiences, because they simply fall asleep too fast. If you want to see what happens on the edge of sleep perform the following procedure.

Procedure
1. Choose a time when it's okay to lose a little sleep – like maybe on a weekend. The best time would be when your body is very tired, but your mind feels fairly alert.
2. Get into bed as usual, but make sure you are on your back. Get very comfortable.
3. Now lift one hand and forearm off the bed, bending at the elbow and leaving the elbow and upper arm on the bed. *(If you do it right, you can easily keep your arm propped up that way for a long time.)*
4. Try to let your body fall asleep, while keeping your mind focused on what you are experiencing.
5. Just as you fall asleep, your arm will start to fall.
6. If the experiment works right, you will wake up enough to pull it back into balance. For some people, this little trick lets them prolong hypnogogia, and stay alert enough to remember the experience.
7. *(This experiment may not work for everyone)*

SNORING

Snoring occurs when the flow of air through the mouth and nose is physically obstructed. Air flow can be obstructed by a combination of factors.

Observation Procedure
1. Do you know anyone who snores?
2. If so, listen carefully to their snoring sounds.
3. Is the snoring very regular?

4. Are there pauses, snorts, or gasps?
5. Are there periods when he or she seem as if they are not breathing? If so, this could be an indication of sleep apnea.

SLEEP LATENCY

In this exercise you will figure out how long it takes you to fall asleep.

Materials
- Spoon
- 1 standard-sized 18" x 26" x 1" Aluminum Bun Pan/Sheet Pan
- Timer

1. Write down the time you get into bed.
2. When you are in bed trying to get to sleep, hold a metal spoon over an Aluminum Bun Pan/Sheet Pan on the floor.
3. When you fall asleep, your muscles will relax and the spoon will fall out of your hand.
4. The noise of the spoon hitting the plate should wake you up.
5. Write down the time you woke up.
6. The difference between the time you got into bed and time you woke up is your sleep latency.

(NOTE: If the spoon misses the plate, you may not hear a noise signaling you to wake up.)

SLEEP LOG

A sleep log is used to study sleep by keeping a record of your sleep behavior and the dreams that you have each night. It is best to write down the dreams immediately when you wake up because the events and details of the dreams fade with time. So keep your lab manual and a pencil next to your bed when you perform this exercise. After a few nights of sleep and dreams, you will get better at remember what happened in the dreams. You will have to make 7 copies of the **"Dream Journal"** page.

When you are logging your dreams, ask yourself these questions.
1. Are your dreams in color?
2. Do you have a "sense of time" in your dream?
3. What emotions did you have during your dream?
4. How many different dreams can you remember in one night?
5. Do the same people, events or places reoccur in different dreams?
6. Do some events that happened during the day appear in your dreams?
7. If you think about something before going to sleep, does this "something" appear in your dreams?
8. Does watching a movie or a TV show influence what you dream about?
9. Does eating certain food influence what you dream about?
10. Does your mood affect what you dream about? If you are happy, do you dream about different things than if you were sad?
11. Are dreams on weekdays different than dreams on the weekends?
12. Does the time of year influence what you dream about?
13. Does the time you go to sleep influence what you dream about?
14. Are nighttime dreams different from dreams you have if you take a nap (or fall asleep during class)?
15. Are dreams different when you are sick?
16. Are dreams different when you take medicine?
17. Do you remember dreams you have had in the past? How long ago?

18. Do you have the same dreams more than once?
19. Are your dreams similar to the dreams of other people in the class? Compare some of your notes.
20. Can you remember your dreams better when you wake up by yourself or when you wake up with an alarm clock?

Dream Journal
(Keep track of your dreams)

Date:_____ Day of the Week_____
Time to Bed_____ Time Awake_____

Describe your dream(s) that you had in the spaces below. Use a separate sheet of paper to write down the dreams you had each night. Include as much detail as you can remember.

Dream 1
_____
_____
_____
_____
_____
_____
_____
_____

Dream 2
_____
_____
_____
_____
_____
_____
_____
_____

Dream 3
_____
_____
_____
_____
_____
_____
_____
_____

Sleep Journal
(Record your sleep behavior)

	Time to bed at night	Time awake in the morning	Duration and number of times awakened during the night	Total Sleep Time	In the morning, how did you feel?	Did you remember any dreams?	What did you do one hour before going to sleep?
Day _____ Date _____	____ pm ____ am	____ am	____ minutes ____ times	____ hours ____ minutes	__ tired __ refreshed	__ yes __ no	
Day _____ Date _____	____ pm ____ am	____ am	____ minutes ____ times	____ hours ____ minutes	__ tired __ refreshed	__ yes __ no	
Day _____ Date _____	____ pm ____ am	____ am	____ minutes ____ times	____ hours ____ minutes	__ tired __ refreshed	__ yes __ no	
Day _____ Date _____	____ pm ____ am	____ am	____ minutes ____ times	____ hours ____ minutes	__ tired __ refreshed	__ yes __ no	
Day _____ Date _____	____ pm ____ am	____ am	____ minutes ____ times	____ hours ____ minutes	__ tired __ refreshed	__ yes __ no	
Day _____ Date _____	____ pm ____ am	____ am	____ minutes ____ times	____ hours ____ minutes	__ tired __ refreshed	__ yes __ no	
Day _____ Date _____	____ pm ____ am	____ am	____ minutes ____ times	____ hours ____ minutes	__ tired __ refreshed	__ yes __ no	

EXERCISE 9.21
IDENTFYING PARTS OF THE EAR

Purpose of exercise: To identify parts of the human ear.

Now click on *Exercise 9.21* within your online platform or enter the address below into your web browser:
https://www.biodigital.com/

- *(This is a free site, but you will need to sign up with your name and a validated email address. You can use your personal email address or school email address. MAKE SURE TO CREATE A PASSWORD YOU CAN REMEMBER. I WOULD SUGGEST WRITING IT DOWN AND KEEPING IT IN A SAFE PLACE. YOU WILL USE IT AGAIN.)*

Click **SIGN UP**. Once you have provided the appropriate information, you are now able to access BioDigital's website content. Click **LOG IN** using the email and password you created. Then click **SIGN IN**. On the left of your screen choose from the systems listed.

Please click on the square boxes next to this icon. . Now click on the **Anatomy By Systems** icon. Click on the following image(s) with caption. **Male Auditory System, Female Auditory System.** View the structures associated with this system.

Color the parts of the ear.

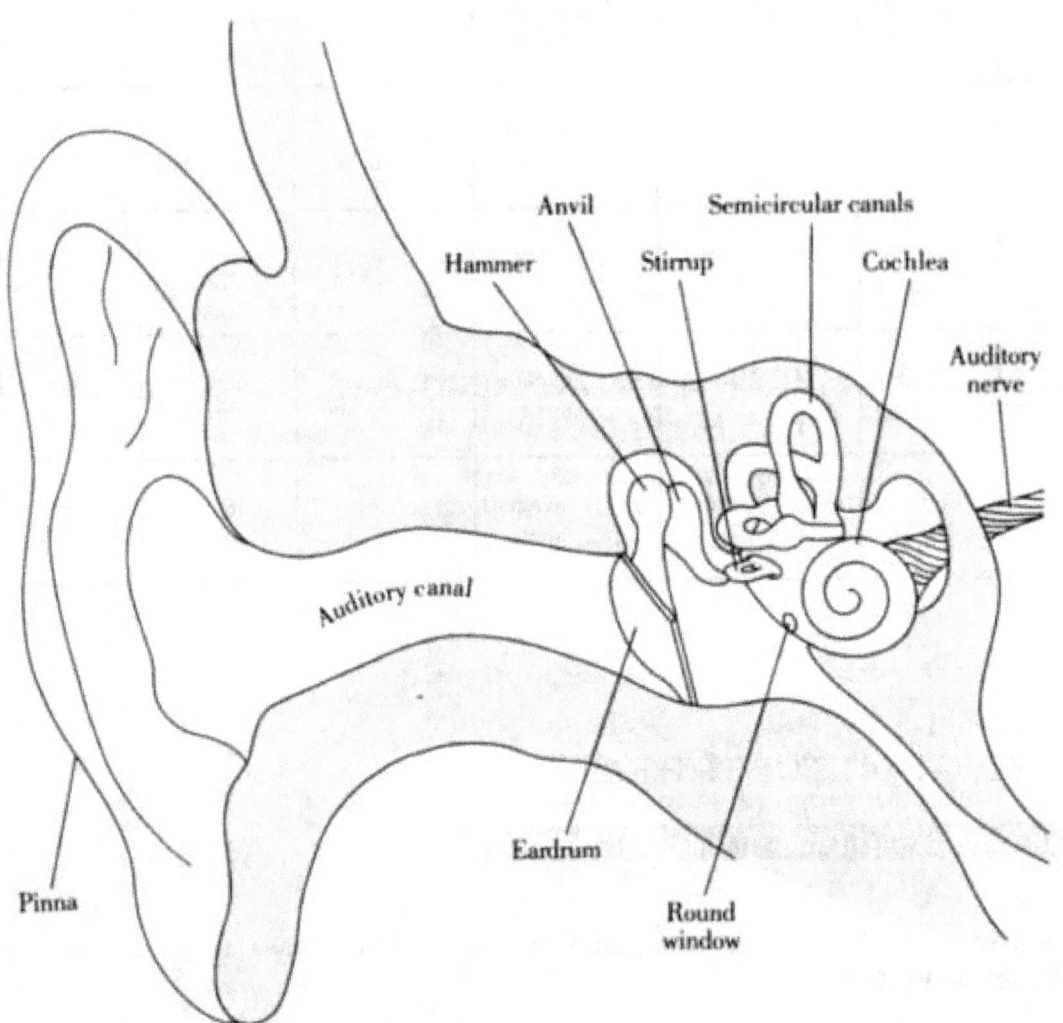

EXERCISE 9.22
REVIEW QUESTIONS

1. The spinal cord is located in the _____ canal and is enclosed by the membranes called _____.
2. In a cross-section of the spinal cord, the _____ matter is interior and is surrounded by the _____ matter.
3. The _____ tracts of the spinal cord carry _____ impulses toward the brain.
4. The _____ tracts of the spinal cord carry _____ impulses away from the brain.
5. The dorsal root of a spinal nerve contains _____ neurons.
6. The ventral root of a spinal nerve contains _____ neurons.
7. The number of pairs of spinal nerves, in order, from cervical down, is _____, _____, _____, _____, and _____.
8. The part of a reflex arc that detects a stimulus is the _____.
9. The part of a reflex arc that performs the reflex is the _____.
10. The part of a reflex arc that carries impulses to the CNS is the _____.
11. The part of a reflex arc that carries impulses away from the CNS is the _____.
12. A reflex is defined as a(n) _____ response to a(n) _____.
13. The patellar reflex is an example of a(n) _____ reflex, and the part of the CNS directly involved is the _____.
14. The part of the brain that regulates blood pressure is the _____.
15. The parts of the brain that regulate breathing are the _____ and _____.
16. The part of the brain with centers for visual and auditory reflexes is the _____.
17. The part of the brain that regulates muscle tone is the _____.
18. The part of the brain that regulates body temperature is the _____.
19. The part of the brain that regulates secretions of the anterior pituitary gland is the _____.
20. The part of the brain that produces the hormones that are stored in the posterior pituitary gland is the _____.
21. The part of the brain that integrates sensations to relay to the cerebrum is the _____.
22. The part of the brain that suppresses unimportant sensations is the _____.
23. The part of the brain that is essential for the formation of memories is the _____.
24. The _____ lobes of the cerebrum initiate voluntary movement.
25. The _____ lobes of the cerebrum feel the cutaneous senses.
26. The _____ lobes of the cerebrum contain the auditory areas.
27. The _____ lobes of the cerebrum contain the visual area.
28. The _____ lobes of the cerebrum contain areas for taste and smell.
29. The cerebral hemispheres are connected by the _____.
30. The meninges, in order from the surface of the brain outward, are the _____, the _____, and the _____.
31. Cerebrospinal fluid is found in the subarachnoid space and ventricles of the _____.
32. The _____ division of the autonomic nervous system dilates the bronchioles and increases heart rate.

Special Senses

33. The part of a sensory pathway that detects a change is the _____.
34. The layer of the eyeball that forms the white of the eye is the _____.
35. The most anterior part of the sclera of the eye is the _____.
36. The cornea differs from the rest of the sclera in that the cornea is _____.

37. Tears contain _____ to inhibit the growth of bacteria.
38. The _____ glands produce tears.
39. The _____ layer of the eyeball absorbs light within the eye to prevent glare.
40. The part of the eye that contains the receptors for vision is the _____.
41. The eyeball is moved up and down or side to side by the _____.
42. The _____ help the eyelids keep dust out of the eye.
43. The size of the pupil of the eye is regulated by the _____.
44. The part of the eye that contracts to change the shape of the lens is the _____.
45. The anterior cavity of the eye contains _____ and the posterior cavity contains _____.
46. Aqueous humor is reabsorbed back into the blood through the _____.
47. The photoreceptors in the retina are the _____ and _____.
48. The optic nerve passes through the eyeball at a site called the _____.
49. The first part of the eye that refracts light rays is the _____.
50. The only part of the eye that adjusts to focus light rays is the _____.
51. The receptors for hearing are located in the _____ of the inner ear.
52. The receptors that detect motion are located in the _____ of the inner ear.
53. The first part of the ear that vibrates in response to sound waves is the _____.
54. The auditory bones, in the order in which they vibrate, are the _____, _____, and _____.
55. The cranial nerve pair concerned with hearing is the _____.
56. The auditory areas are located in the _____ lobes of the _____.